Third Edition

Quantum Physics:
an overview of a weird world

*A primer on the conceptual foundations
of quantum physics*

Marco Masi Ph.D.

Third Edition 2019

Cover image: 3D probability density of hydrogen-like atomic orbitals. (By Geek3 - https://commons.wikimedia.org/w/index.php?curid=69035277)

Contents

Introduction

Nowadays, there are many popular books, textbooks, and online courses teaching quantum physics (or 'quantum mechanics' or 'quantum theory' as usually referred to interchangeably). For undergraduates and graduates, many well-written textbooks are available that elucidate every conceivable aspect of quantum mechanics (QM). Additionally, there are many accessible books at the popular science level. In most cases, however, these materials take one of the following two attitudes and perspectives towards the subject.

On one side, there are treatises that require the reader to have a solid background in mathematics. One cannot go very far with university-level textbooks if the student does not grasp the basics of, at a minimum, calculus, and linear algebra, as well as many concepts from classical mechanics. Moreover, most of these sources are concerned primarily with the calculations and purely technical aspects of QM and pay only limited attention to a deeper understanding of the more foundational implications.

The mathematical complexity of these sources is usually so time- and energy-consuming (albeit, in certain respects, fascinating) that the effort to reduce everything into a formal language no longer allows for additional conceptual investigation. Among physics graduates and even well-established physicists, the "shut up and calculate" working approach reigns, which looks with skepticism at foundational issues. Contrary to popular belief, most physicists are not at all concerned with the philosophical foundations of quantum theory (QT), but are much more focused on doing calculations—the process of which can obscure and hide the deeper meaning and nature of the physical phenomena that the math describes. For this reason, this book's focus on the conceptual foundations of QM and its resulting implications may make it of interest to those who have already studied physics, as it provides a conceptual understanding of QM that goes beyond even that which most physicists embrace. Yet, at the same time, this book does not need any technical background except for some basics of algebra, trigonometry and calculus. It might be easier for those having already some math background but a mathematical appendix is furnished for those who need a reminder.

On the other side, at a more popular level, an extensive body of literature and online videos is available nowadays for the individual who does not have a scientific background and who does not intend to become a professional physicist. Unfortunately, however, most of this popular science dwells on the surface or, worse, consists of pseudo-scientific ruminations by which the author seeks to impress the audience with misleading (or even false) statements. This is especially true when it comes to the philosophical aspects of the foundations of QP. Most of these books, some written by qualified and world-known physicists, reduce the concepts to such a low level that, paradoxically, they become unclear and difficult to understand simply because they treat their audience like children. Moreover, many best-selling books and science documentaries promote highly dubious ideas and speculations backed by no evidence. Sometimes these ideas can't be verified experimentally, not even in principle, and have not worked out. This seemingly pedagogical approach of many popular and vulgarized versions of quantum physics (QP) has done more harm than good and has misled the audience regarding the real nature of the quantum world. Another aspect which plagues most popular introductions to QP is the author's desire to explain at length their own perspectives, theories, and conjectures on a subject they are working on or that reflect predominantly their personal preferences and intellectual inclinations. In doing so, they focus too much on details and specific topics which are not necessarily relevant for a first impact primer on QP. Thus, they no longer furnish the promised necessary general basics that the reader sought.

Therefore, there is a missing link between the professional, fully mathematized exposition of the principles of QM and that which is accessible to a broader audience but which ultimately leaves them with doubtful insights. The net result is that, nowadays, we have a scientific community that is busy with a plethora of calculations but that has lost contact with a more profound understanding of its foundations. On the other hand, the public has, in its mind, a cartoon version—or, at best, only a very partial understanding—of one of the most fascinating subjects of modern science.

This leads me to the point that this book is not simply a series of lectures; it means much more to me. It is a long-term undertaking which aims to demystify QP, freeing it from the decades-long existence of disinformation to which the media has subjected their audience. It is an attempt to uplift people's awareness of a topic which has been vulgarized and abused for at least a couple decades of low-level popularization.

Despite being dedicated to non-experts, this book is also an attempt to encourage undergraduate students to look beyond the 'shut up and calculate' approach that still permeates our universities and colleges. It invites them to deepen the conceptual foundations of QP and its philosophical implications—something which academia and society, anchored in a more pragmatic and utilitarian approach, unfortunately rejects and ignores. In particular, I recommend this course to undergraduate and graduate students of physics, philosophy, the history and philosophy of science, engineers or students in the IT branch who, because of the nature of their faculties, have already some analytic background and who would like to focus on the conceptual foundations of QP. For professional physicists as well, though they already have a technical and mathematical background, it is unfortunately still rare and unlikely that their university or college has prepared them to tackle the foundational issues of QP.

There is still a divide, a large split between scientists and philosophers, a split between the purely formal and technical approach of mathematicians and physicists (which we will avoid here as far as possible), and the approach of the philosopher who desires to look further beyond the math at more conceptual issues. This course seeks to close that gap. It also works towards a proposal for approaching science in general from a different perspective then schools, colleges, and universities presently pursue. It would be great if students could convince their departments and faculties of philosophy or physics to recognize it as an integrating part of their curriculum.

I would like to bridge this gap between a merely mathematical exposition and an overly simplistic popular version of QP by presenting an overview of the basic concepts that stand behind non-relativistic QT with an emphasis on experimental and technical facts that should prepare you to correctly understand (first and foremost) its foundations, with particular attention paid to its philosophical implications.

Our journey through the weird world of QP will span from the very first principles, such as the concept of the quantum and the blackbody spectrum introduced by Max Planck at the beginning of the 20th century, to the most recent developments in string theory or quantum cosmology. However, the number and complexity of topics necessary to make the subject accessible to the layman are quite large and intricate. Therefore, it seemed appropriate to separate this overall 'grand vision' into two parts.

The present book deals with all those aspects of QP, especially those concerning the foundational issues, which have nowadays become the building blocks and experimentally established scientific facts. It is the introductory primer everyone requires to obtain a solid background and which essentially dwells on all those conceptual foundations of QP which received empirical validation throughout the 20th century. It is a self-contained book for beginners.

A second upcoming volume will be a presentation of special selected topics dealing with more recent developments in the field and rests on the more speculative outgrowths of QP. It is a more advanced overview dedicated to advanced readers who are willing to put additional effort into dealing with more complicated topics which require more time and intellectual dedication.

Broadly speaking, one might say that the first volume is a self-contained introduction to QP which tells us what we know for sure about the physical world, while the second volume describes primarily (with the exception of the standard model of particle physics) what modern physics is still working on. In this later edition it will provide the most recent advances and developments in the field. (If, at the time you are reading this, the second book has not yet been published, you can subscribe to an alert list which will inform you about when this will happen. Simply send a note to marco.masi@gmail.com.)

The philosophical implications and the correct didactical approach to QP remain a matter of personal preference and subjective interpretation. There are countless interpretations of QP (which will be one of the selected topics edition). One might say that there are as many interpretations of QM as there are physicists, and you should be aware that physicists can't agree on what the quantum world's ontology is really about. This debate will not be resolved soon. As some surveys reveal, many physicists don't even care or have an opinion. [1] Most are busy conducting experiments and doing the math that describes quite well the phenomena; they tend to refrain from interpretational aspects because they know this to be an 'intellectual minefield'. It is easier to say what the QP is not than what it is, and several hundred pages would be required to do justice to all these lines of thought. What you can expect to gain from this introduction are the tools and the basic technical and experimental facts which will allow you to frame your own understanding and hypothesis, and then embrace the interpretation that you feel most in line with.

However, this led the author to a dilemma. Because there are so many possible interpretations, it is impossible to write a book on QP that takes a perfectly objective and 'agnostic' stance while not letting one's own point of view and interpretations slip through. The quantum world is so weird that everyone sees something different in it. I will, therefore, not hide my preference for a QT that is interpreted as non-local, indeterministic, without hidden variables, and that does not admit counterfactual definiteness as a logical inference tool. (If this statement confuses you, don't worry; we will discuss this in detail.) This might not satisfy those who support the opposite point of view in which QM is envisaged as a deterministic theory with hidden variables. Such is especially true of those who defend the de Broglie-Bohm pilot wave theory or Everett's many world interpretation. However, that I will be looking upon the facts from the perspective of a 'non-determinist' (just as every other physicist does from his or her own perspective based on subjective preferences, whether they want to admit it or not) should not be of much concern, as what we will first and foremost lay out are the facts, the empiric and experimental phenomena, and, as far as an introductory document like this allows, the different perspectives and theories of the different scientists involved. There is no loss of continuity. Once these are known (this document is devoted primarily to providing you with this knowledge), everyone can take their own personal standpoint and embrace whatever interpretation they desire.

Some remarks about the necessary background: While we will go as far as possible without complex mathematical calculations, we will also keep things as rigorous as possible, as far as these non-mathematical limits allow. However, some material inevitably requires a bit of algebra. Those who still have in their minds the standard high school math might have some advantage, but you do not need particular mathematical skills. Only an elementary understanding is sufficient of the Pythagorean theorem, square root, exponential, sin/cos functions, Cartesian coordinates, vectors, the intuitive concept of a derivative, basic notion of a complex number, etc. For those few parts where some mathematical description is unavoidable, a quick mathematical primer has been added in the appendix so that no one is left behind. In the appendix you will also find the list of acronyms used throughout this book.

Let us briefly take a look at the main topics that will be covered.

In the first two parts, we will discuss the birth and foundations of QM. Here, we will begin with a historical overview from the very first conceptual foundation. It starts with the old controversy regarding whether light is a wave or made of particles. This question seemed to receive a definite answer with Young's double slit experiment, and yet it had to be reviewed later when QP was born. Historically, the birth of QM is identified with the intuition of Max Planck that explained the so-called 'blackbody radiation' by postulating that energy manifests only in quantized forms. This will be connected to the description of that unexpected phenomenon, the photoelectric effect, which led Einstein to work with the modern notion of the ‚photon' (that is, the light particle). Bohr's atom, the famous uncertainty principle of Heisenberg, and the even more famous wave-particle duality, as well as other notions, will follow as ingredients necessary for understanding the basis of modern QT.

The following part III will address some important technicalities that are nonetheless interesting. First, the quantum world of spinning particles is a short but intense description of notions like the spin of particles, with a special emphasis on the Stern-Gerlach (SG) experiments. It is not intended to provide a complete and comprehensive view of all the developments of QT in the last century, as this would require at least twice, if not three times, as much space and would lead us beyond the introductory character of this book. It is, nevertheless, a selection of those topics and findings that are especially interesting for their philosophical implications and a deeper understanding of quantum phenomena. The parts entitled 'quantum ubiquity and randomness' and 'quantum entanglement' will focus on the aspects of QT which sparked so many discussions and speculations among scientists and philosophers in the last decades. The typical example is the famous thought experiment of Schrödinger's cat. Much more will follow, such as the meaning of quantum superposition and entanglement, whether empty space is really empty according to QP, the zero-point energy, the time-energy uncertainty relation, and the quantum tunnel effect.

Further on, we will set the stage with some basics about the deeper significance of the Pauli exclusion principle and will try to provide some answers to seemingly naive questions that receive a clear answer only within QT, such as why matter appears to our sense as 'hard' and stable. We will also review some quite strange but interesting and relevant experiments like those that confirmed the Aharonov-Bohm effect, path integrals and Feynman diagrams and the quantum Zeno effect.

This will lead us to address in 'Bell's legacy' the famous Bell theorem and its related inequality which allows us to discriminate between a local realism and a classical theory, or non-classical faster than light (FTL) theories that have sparked many controversies among physicists and philosophers of science. We will investigate several other key aspects and experiments as well.

What I like to call 'quantum ontology reborn' will follow. We will turn our attention towards some of the most intriguing and amazing modern experiments that were previously impossible to perform due to technical limitations. In fact, modern technological advancements have made it possible, in the last couple of decades, to further test the predictions of QT. For example, the so-called 'which-way', 'interaction free', and 'delayed choice' experiments, could be conceived of before the development of more advanced technology only as ideal thought experiments but can now be tested experimentally. Some concluding remarks will follow.

The aim of this book is not to be an overall comprehensive 'Bible' of QM. Instead, with this bag of knowledge, you will have a solid background in the conceptual foundations and philosophy of QM, which will allow you to discriminate between well-founded statements and pseudo-scientific hype that falls into and out of fashion. I hope this will then encourage you to learn more and proceed further to the second volume in your quest to realize greater knowledge about this fascinating subject.

I.   Quantum prehistory: particles, waves, and light

1. The question of the nature of light

Since the days of ancient Greece, natural philosophers have inquired about the physical nature of light. What exactly is light? Where does it come from? How is it emitted and absorbed? One of the most challenging sets of questions that has kept many bright minds busy to this day are: "what is light made of? Is it made of 'light particles' as water waves are made of water molecules? Or is it a wave? If it is a wave, then a wave made of what and traveling through what? Or is it both a wave and a particle? Or is it something entirely different?" These are questions with almost no answers.

Fig. 1. Johann Wolfgang von Goethe (1749-1832)

Aristotle declared that *"the essence of light is white colour and colours are made up of a mixture of lightness and darkness."* In hindsight, this was not a bad guess. Interestingly, the idea of light being a mixture of lightness and darkness was taken up later by J.W. von Goethe, a German writer, poet and natural philosopher living in the 18th century. Goethe constructed a theory of light that indeed works. He conceived light as not being something fundamental but something that emerges from a 'primordial phenomenon' (from the German 'Urphänomen'). This primordial phenomenon can be observed at the boundary between light/darkness. In this formulation, colors never arise from light alone but as consequence of a dark/light or a black/white contrast at boundaries, edges, and slits. This interesting theory, on its own, could not be disproven and is, in principle, still valid. I would recommend everyone interested in the history and philosophy of science analyze Goethe's approach. It might still have something to teach us as it explains light, darkness, and colors from a qualitative standpoint, and, in some respects, could be considered a complementary understanding to the modern physical quantitative approach.

Goethe's theory of colors, being a purely qualitative theory, cannot make quantitative predictions as a modern quantitative science – like physics – needs to be able to do. Goethe tried to challenge the theory of Sir Isaac Newton, the famous English physicist, mathematician and astronomer, better known for his law of universal gravitation, who was able to separate white light into its coloured components using an optical prism. But more relevant for our purpose was Newton's 'light particle theory'. According to this conception, light is a flux of 'coloured particles'.

At the time, this was not much more than speculation; it was not a real understanding coming from conclusive investigation. In fact, Christian Huygens (1629-1695), a Dutch mathematician, physicist and astronomer, did not share Newton's idea either and argued to the contrary, that light is, in fact, a wave.

Fig. 2. Isaac Newton (1643-1727); Christian Huygens (1629-1695)

For a significant amount of time, there were heated debates regarding which theory was correct. In fact, it was difficult to imagine light as made of particles and, at the same time, being a wave. One theory must be true, but both being true was considered to be a logical impossibility.

In 1803, a groundbreaking discovery settled the issue for over a century. Thomas Young, an English physician, performed a revolutionary experiment, the well-known 'double slit experiment'. He first showed that when light is emitted by a point source and is directed through a card with two tiny pinholes, a colored interference pattern appears on a screen positioned at an appropriate distance on the other side of the card with the pinholes. Later, the same experiment was performed with two slits instead of pinholes. How waves add or subtract each other, resulting in an interference pattern, was something that was already well-known and accurately studied with mechanical waves, typically oscillating pendulums, or water waves. Since light was shown to exhibit the same characteristic behavior in the double slit experiment, this seemed to finally prove the wave theory of light vs. the particle interpretation beyond any reasonable doubt.

Further insight on the nature of light came from the work of James Clerk Maxwell (1831-1879), a Scottish theoretical physicist, who predicted the existence of electromagnetic (EM) waves. In 1864, he wrote his famous Maxwell equations, which are the mathematical foundation for modern electromagnetics. These are a set of four elegant equations that are very general and hold for all EM waves. The connection between light and electricity as magnetic phenomena became clear as an obvious consequence of the fact that electric and magnetic fields propagate with the same speed of light. So, the issue seemed to have been settled once and for all. Light must be an oscillating electric and magnetic field propagating throughout space like a wave. These equations were accepted as undeniable fact.

Fig. 3. Thomas Young (1773-1829); James Clerk Maxwell (1831-1879)

One problem still remained, however: what is the medium through which light is supposed to travel? Water waves, sound waves or material waves use respectively water, air or a solid body as a support through which they can propagate. Without water or air or matter in some form, there can be no transmission of mechanical waves. Since light can travel through empty space, this led naturally to the question of if empty space were made of some subtle, yet unknown 'luminiferous aether' that could function as a medium for EM waves. These ruminations kept the young Albert Einstein busy while he was working in a Swiss patent office and led him to his famous theory of relativity. He posited that there is no aether (without even knowing that this had already been shown experimentally) and drew from that the conclusion that made him world-famous: the theory of special relativity (SR).

The idea that empty space is not made of some immaterial substance that offers light a transmission medium has held until today and no longer causes much debate. However, as we shall see, our conception of light being a wave and not a flux of particles will turn out to be in need of much further and deeper clarification and analysis.

However, before continuing in our historical analysis of the development of the nature of light, it is necessary to introduce an interlude that sets the conceptual foundations we will need throughout the book. Nothing can be said in QP if we do not have a clear and rigorous understanding of how waves and their characteristic interference phenomena are described in physics.

2. Waves and Interference

So, what is interference? It is extremely important that you understand at least intuitively what this is about. Interference is so important because it is the physical effect that stands behind a lot of phenomena in QP. Without the wave-description and the interference of 'wavy' phenomena, almost nothing can be said in QP.

Let us first of all see how light is conceived of as an oscillating EM field. To see what a field in physics is, the example of a force field like that of gravity might be more illustrative since we know that better from our everyday experience. The gravitational field that surrounds a planet like the Earth is a region of space where gravity acts upon material bodies. The moon interacts with the Earth's gravitational field and also the reverse is true; that is, the Earth 'feels' a force originating from the gravitational field of the Moon. Also, other forces acting upon objects exist, namely electric and magnetic forces, the actions of which on the surrounding space can be visualized with the field notion as well. What distinguishes the electric and magnetic field forces from gravity is that the former can be both attractive and repulsive – in contrast to the latter, which can be only attractive, at least in classical physics (CP). Electric and magnetic forces are therefore characterized by a polarity (positive and negative electric charges, north and south magnetic poles).

A detailed description of these fields would require a lengthy elaboration that is not essential for our purposes here. It need only be highlighted that a moving electric charge always produces a magnetic field (say, an electric current in a wire will always create a magnetic field whose lines are concentric to the wire). And, when an electric field changes in time, for instance a periodically oscillating electric field caused by an electric charge oscillating in space, this always induces a corresponding oscillating magnetic field. The reverse is also true: an oscillating magnetic field always gives rise to a corresponding oscillating electric field. The two never exist separately when one or the other varies in time. This is the reason why physicists speak of oscillating EM fields, unifying the concept of electricity and magnetism in a unique and inseparable physical entity, a concept that is central in Maxwell's equations.

This can be visualized as in *Fig. 4*, with an electric and magnetic component oscillating in time and traveling in space up to the speed of light. Electric and magnetic forces have a direction and a strength, and for this reason are represented graphically as arrows, the lengths of which indicate the magnitude of the field in each point of space and its direction in the orientation of the acting force.

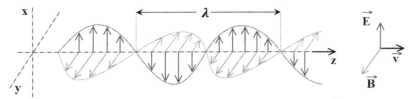

Fig. 4 The propagating EM wave.

Mathematically, one represents these 'arrows' by vectors (see also the mathematical review in Appendix A Ib for more). In *Fig. 4*, as has become a convention, the vector of the electric force is labeled 'E' and that of the magnetic one is labeled 'B'. Formally, each component of an EM field can be represented by a function of these vectors changing in time while moving along the z-direction. The oscillation plane of the electric field (in the figure, the vertical E-B plane) determines the so-called *'polarization plane'* of the wave. Remember always that two waves can have the same wavelength (or frequency) and the same direction of propagation but different *'polarizations'*. The orientation of this plane determines the orientation of the polarization. In this case we have vertical polarization but, of course, any other orientation is possible, as the horizontal one or any other in between (more on this later).

It can be shown that, since the electric field is a result of a changing magnetic field in time, and vice versa, the electric vector must always be perpendicular to the magnetic one. And, as you can see, to this oscillatory field we can associate a wavelength λ ('lambda'), which is the measure of the 'wave's size' or, more precisely, the length between two identical field vectors with the same amplitude and direction, or to put it more simply, twice the distance between two points where the field is instantaneously zero. Mathematically it is just the speed of propagation of the wave c, divided by the number of cycles, that is, its frequency ν ('nu'). So, in general:

$$\lambda = \frac{c}{\nu}.$$

Now, how do we know that light is, or behaves like, a wave? Well, it is because of interference phenomena. To illustrate this, let us take two waves which have the same amplitude and are 'in phase', as in Fig. 5a (to keep it simple, we assume the polarization to be the same as well). Here, the peak or valley values (say, of the electric field) of one wave are coincidental in time and space with the peak and valley values of another wave. In this case we will have a peak-to-peak superposition, and they will sum up to form a single wave with larger amplitude. This is called *'constructive interference'*.

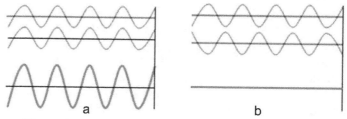

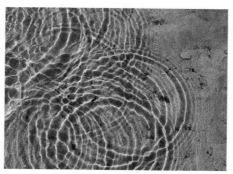

Fig. 5 a) Constructive interference (waves in phase) and
b) destructive interference (waves in anti-phase).

In the case of the opposite situation of Fig. 5b—that of two overlapping waves with the same amplitude, but in anti-phase—then one has a peak-to-valley superposition, whereby the two waves cancel out each other and the resulting signal will be zero. This is called 'destructive interference'.

In a situation which demonstrates the in-between of constructive and destructive interference, where two or more waves with different amplitudes and phases come together and interfere to produce a somewhat fuzzy superposition of waves, you will of course get a less regular and ordered signal.

A typical example of this is what you observe when you throw two or more stones into a pond (Fig. 6). You get a pattern where you have some places where the waves add up, others where they subtract from each other, and still others where they interfere in other ways. We will look at another example of this soon.

Fig. 6 Interference phenomena in water.

We are now in a position to properly describe interference phenomena. Let us address the question of precisely how intersecting waves add to or subtract from each other.

Using the Greek symbol for the amplitude of a wave at position x at time t, which is written $\psi(x, t)$ generally (read: 'psai of x and t'), let us add two waves with functions $\psi_1(x, t)$ and $\psi_2(x, t)$ (for those who are not familiar with mathematical functions, see Appendix A Ic for a brief introduction on how functions are used to describe waves). These could be representative of the amplitude of the electric or magnetic field of two light beams. To keep things simple let us consider the case of two waves with a different frequency but with the same amplitude, as in Fig. 7.

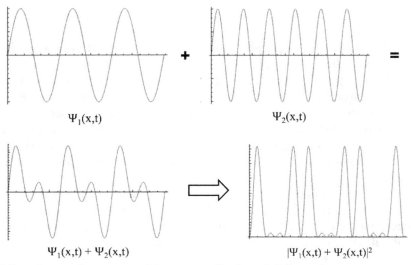

Fig. 7 *Interference of two waves with same amplitude and different frequency.*

We won't go into the calculations, but one can show how, with the above described mathematical structure, the interference resulting from the superposition of both fields looks like the resulting signal of the lower left-hand depiction in Fig. 7. This function still has negative values and therefore would make no sense as a measure of the intensity of the light resulting from this overlap. A negative intensity would also be a meaningless concept in physics. Therefore, what one does is to square these values, since the square of any number, be it positive or negative, always gives a positive value. And this is what is done in physics to obtain a light intensity measurement from the EM oscillatory field it represents. So, taking the squared modulus of the sum of these fields leads to the lower right-hand side interference pattern. The squared modulus is used because the two functions are usually represented by complex numbers (see Appendix A Id).

As you can see, for this particular case, one obtains two intensity maxima, two of them small and the other two much larger, which are arranged alternatively along the horizontal x-axis. Notice that this pattern results only for a particular time t. In reality these peaks oscillate, changing their intensity with passing time. But for light, this oscillatory movement is so extremely fast that it is undetectable—even for the most advanced modern measuring devices (the frequency of yellow light is about $5 \cdot 10^{14}$ Hz, which is 500,000 billion oscillations per second). Therefore, since the oscillatory phenomenon of the electric and magnetic fields of light occur on a time scale well beyond the measurable domain, for most practical applications one can just use the average intensity, which turns out to be a very good approximation. That's why we will frequently skip the time parameter in the wave description and simply write them as waves with their spatial dependence without the time parameter, as $\psi(x)$.

The important message to relay here is that of how waves interfere and that their intensity results in constructive and destructive interference patterns. You should develop at least a clear intuitive understanding of the above described interference phenomenon, because it is something we will encounter over and over again.

3. Young's double slit experiment

Now let us see how this concept can be applied to the special case of the double slit experiment, which we will consider frequently throughout this book. We will start with a brief introduction.

Please consider that this is one of the few sections that indulges a bit in mathematical description, because it is necessary to state clearly what happens with the interference phenomenon of waves. If you are no longer acquainted with the mid-school trigonometric functions, exponents, and square roots, please refer first to the appendix where these simple concepts are elucidated and applied to the description of waves.

The experimental setup is that of a wavefront which comes from the left and is diffracted at two small slits, as shown in Fig. 8a.

In general, such a wavefront could belong to water, or light, or sound, etc. Interference phenomena are a common characteristic of any kind of wave, not only of EM waves. Ideally, one should use a highly monochromatic source of light, like that of a laser, in order to obtain a single wavelength front (of course, this was not feasible during the time of Young; he just used a candle, and with optical filters it is possible to obtain interference fringes thereby, even if they appear in several colors).

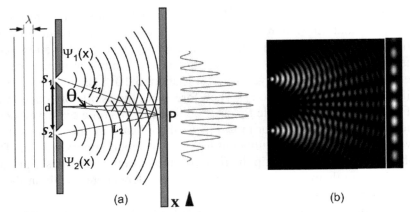

Fig. 8 The double slit experiment: a) Diagram b) Visible wavefront

More specifically, 'small' slits are about the size of the wavelength of the incoming wave. If the wave were to have a much larger wavelength than the size of the slits, it would simply bounce back or be absorbed by the wall as if there were no pinholes or slits at all. If, on the other hand, the wavelength of the incoming wave were much smaller than the slits, then it would experience almost no diffraction.

But if the wavelength is comparable with the slit size, then we will observe a situation like that shown in Fig. 8b. After the slits' apertures, and because of the diffraction of the transverse plane wavefront, two concentric outgoing wavefronts form (in a three-dimensional volume, these would be a pair of spherical wavefronts).

Diffraction is characteristic of every wave phenomenon and occurs whenever a wave encounters an obstacle of a size comparable to its wavelength. Every time light passes by an edge or a little aperture that is comparable in size to its wavelength, it gets diffracted—that is, deflected, bent and deformed. You may have experienced this in your daily life. If you look at a source of light from the edges of an object, such as from a cusp angle or through a tiny pinhole, you will see how the contour, the lineaments of the figure of the object, get a bit distorted and smeared. This is what causes unsharp shadows.

Notice how the two outcoming waves have an intensity which has an angular dependence (more intense 'head on' than seen from the sides). These travel forward in the direction of the screen where they will interfere. The phase difference between the two waves—that one coming from one slit and the other from the other slit—is determined by the length difference of the two lines traced from the two slits S_1 and S_2 and meeting at P. In fact, it is the difference in the paths that the two waves experience in reaching the screen at that point which is responsible for the interference phenomenon.

This will form the characteristic interference fringes along the detection screen (in *Fig. 8* along the vertical x-direction). It is this phase difference which varies along the detection screen, which determines whether we will have constructive or destructive interference or something in between. The bright bands are, of course, the places where we have constructive interference, whereas the darker lines represent the destructive one. Overall, the interference pattern depends on the wavelength of the light, the slits' aperture size, their separation and the distance between the slits and the screen.

Notice also that the interference occurs everywhere, not only on the screen. The screen serves the sole purpose of allowing for its detection and making it visible, but waves overlap, sum or subtract each other everywhere in space, as Fig. 8 b shows.

Let us put this into a bit more rigorous mathematical setting. In the case of the double slit, the two light beams emerging from the respective slits can be described by the two waves $\psi_1(x)$ and $\psi_2(x)$. The angular distribution of the light signal on the detection screen is given by θ, the angle between a horizontal line centered between the two slits and that connecting at some point P on the detection screen.

We assume the two slits, S_1 and S_2, have the same aperture and shape and, therefore, the two emerging waves also have the same angular intensity. (Again, keep in mind that the intensity depends on the angle.) However, the two waves will reach the detection screen in P along two paths having different lengths. This is because the distance L_1 traveled by wave ψ_1 from S_1 to P will, in general, be different from the distance L_2 traveled by wave ψ_2 from S_2 to the same point P. Therefore, even though the two waves have exactly the same wavelength λ and amplitude ψ_0, they will nevertheless reach an observer in P along two different paths with phase ϕ_1 and ϕ_2. The relative phase difference $\delta\phi = |\phi_2 - \phi_1|$, will be determined by the absolute length difference of the two paths $\delta L = |L_2 - L_1|$, which ultimately depends on the position of point P on the screen along the vertical x-axis or, equivalently, by the detection angle θ. This is a major point that must be considered accurately and that we will encounter repeatedly.

Now, if one calculates the complex squared modulus, that is the overall intensity appearing on the screen, and taking into account the phase difference between the two interfering waves in P in dependence on the detection angle θ, one gets the following result (the proof is a bit boring algebra that would go beyond the scope of this introductory approach; however, the interested reader can find it in appendix at the end of section A Id):

$$|\psi_1(x) + \psi_2(x)|^2 =$$

$$= |\psi_1(x)|^2 + |\psi_2(x)|^2 + 2\,Re[\psi_1^*(x)\cdot\psi_2(x)]$$

$$= I_1(x) + I_2(x) + 2\sqrt{I_1(x)\cdot I_2(x)}\cdot cos(\delta\phi).\ \textit{Eq. 1}$$

The resultant interference signal on the screen has three terms: the two waves squared modulus plus twice the real part of the product of one conjugate wavefunction times the other one. Because the squared modulus of wavefunctions represents their intensities, I_1 and I_2, we can set $|\psi_1(x)|^2 = I_1(x)$ as also $|\psi_2(x)|^2 = I_2(x)$. However, there is also the third term which emerges from this calculation and which makes clear that, in general, the intensity at a point in space resulting from two interfering waves is *not* the sum of their intensities. The third extra term is the so-called *'interference term'*. According to the value $\delta\phi$ of the relative phase shift between the two beams, one has constructive interference (the interference term is positive when $\delta\phi$ is an even multiple of 2π and adds up) or destructive interference (the interference term is negative when $\delta\phi$ is an odd multiple of π and subtracts). This interference term modulates the intensity of the signal according to the observation angle. Or, in other words, if we move a detector that measures the intensity along the x-direction at each point, the intensity will oscillate up and down according to the peaks and valleys of the $cos(\delta\phi)$ function, which adds or subtracts an amount $2\sqrt{I_1 I_2}$ from the term $I_1 + I_2$. When a photograph is taken (say, with a ccd-camera), each pixel will display a fixed signal intensity (twice the intensity of the single slit light beam) plus or minus an intensity which changes along the vertical axis. Maximum intensity and minimum intensity appear due to constructive or destructive interference, respectively. (The white interference fringes on a black background or, displaying it in a more technical manner, its mathematical reconstruction is usually depicted with a curve like the function graph in Fig. 8 a.) Whereas, the intensity of the fringes decreases with the angle because of the angular dependence of the two outcoming beams (as shown in Fig. 8 b), that is, they are damped out by a *'diffraction envelope'* (for sake of simplicity this latter aspect has not been taken into account in Eq. 1, but the interested reader can graph fringes, as also shown in the figures, just by multiplying it with an exponential damping factor such as $e^{-(\delta\varphi/10)^2}$).

This interference term is a quite frequent manifestation in QM where particles overlap or interact with each other and makes it appearance wherever one deals with waves having constant relative phases. It is an example of a system in *'coherent state'*. Would the relative phase between the two beams not be constant (that is, $\delta\phi$ vary randomly in time, such as in the case of the overlap of EM waves of two different light sources) this interference term would be zero (the randomicity would 'flatten it out', so to speak) and only the diffraction effects would be present and one speaks of an '*incoherent state*'. Later we shall see how the state of a quantum systems can loose its coherence (for example due to environmental interaction), then one speaks of '*quantum decoherence*'.

So, finally, the important point we must understand is why and how we discern between a wave and a particle. A wave will always show up with an interference phenomenon, and the double slit experiment is the most classical device which makes this clear. Newton firmly believed that light is made up of particles. But later, in 1801, Young's double slit experiment (like many other experimental facts showing interference phenomena) suggested unequivocally that light behaves like a wave.

If light is made of particles, interference could not become visible. We can speak of interference of waves, but there is no meaning in saying that two particles interfere with each other. From the standpoint of CP, and that of our everyday experience, it would make even less sense to speak of a single particle traversing both slits at the same time and interfering with itself like a wave once it hits a screen. Or does it have a meaning? We shall see how subtle Nature can be. QT will teach us a very interesting and deep lesson.

II. The birth and foundations of quantum physics

We have seen that light—or, more generally, EM interactions—seems to behave like waves, not as tiny particles, that is as point-like particles. This was the state-of-the-art knowledge until the end of the 19th century. Newton's light-particle theory was degraded to a historical curiosity. But, beginning from the 20th century, as time went by, new experiments and the great insights of many physicists indicated that this could not be the whole story. The particle nature of light struck back and continued doing so for about three decades.

1. The blackbody radiation, the ultraviolet catastrophe and the quantum

At about the end of the 19th century, physicists had to tackle a very ugly and uncomfortable problem. The question was: Why and according to which principles do hot bodies radiate heat? The answer that CP stubbornly gave was unacceptable: Every object which is not frozen at a chilling absolute zero temperature must radiate an infinite amount of heat in the form of EM waves. For example, according to pre-quantum physics, which rests on classical premises, even a block of ice which is not frozen down to a temperature of -273,15°C—that is, say, even only one degree above this limit—must emit infinite energy and destroy the entire Universe in a catastrophic explosion. Every attempt of the best minds of the time led them to calculations and formulas for the emitted energy of a body at a finite but non-zero temperature, which inevitably always diverged. That is, it showed to tend to infinity. For a long time, there seemed to be no way around this absurd conclusion.

Such a paradox is what the famous philosopher of science, Thomas Kuhn, called an 'anomaly', that is, *"a violation of the 'paradigm-induced' expectations that govern normal science"*. Or, in simpler terms, the current theories predict something very different (eventually, even paradoxical) which does not match with reality and which seems to make no sense.

The solution to this problem marked, historically, the advent of modern QT in the last century. It was German physicist Max Planck who, in about 1900, beautifully resolved the enigma. For this reason, he considered the father of QP. Let us analyze in more detail what this was all about.

Consider very hot magma from a volcano, molten stones and rocks at high temperatures. As is well-known, above some temperatures, typically over several hundreds of degrees Celsius, rocks begin to melt and become reddish. They emit red color radiation, which means that they emit EM waves at a wavelength (or frequency) visible to the human eye. The emitted EM waves, however, become invisible to us when the same rocks have a lower temperature, say, about 20° or 27° Celsius (293 or 300 Kelvin), the so-called 'room temperatures'. This makes it clear that a relation must exist between the temperature of a body and the amount and frequency of the EM radiation it emits, which means the color it displays.

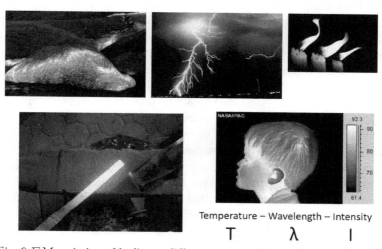

Temperature – Wavelength – Intensity

T   λ   I

Fig. 9 EM emission of bodies at different temperatures and wavelengths.

Something similar can be said for lightning or the light emitted by candles. What happens, in this case, is that an electric discharge heats up the air extremely quickly to thousands of degrees or, like in the case of a candle, a chemical reaction heats up the surrounding air. Here, we are therefore no longer considering a solid or liquid body, like in the case of the solid or molten rocks, but are instead dealing with a gas. The color of lightning is typically very white or bluish, but not red as in the case of lava, contrary to what we would say for the flame of a candle. However, at room temperatures, as everyone knows, air does not emit visible light. Yet it does so at higher temperatures, eventually with very high intensity and sometimes even destructively.

We have a different substance, with different temperatures, but, again, we may ask whether the same relation between the colors—that is, the wavelengths of the emitted EM waves—and the temperature still holds.

Another example is the case of a very hot metal. At room temperatures, it does not emit visible radiation (we will not be able to see it in the darkness), but if we heat up a piece of iron to several hundreds of degrees, it will become visible in the dark, emitting first a reddish radiation. An increase in temperature will cause it to become bluish and, finally, to emit white radiation.

But all this does not happen only in the visible domain of light. The truth is that every object, and even our own bodies, also emits radiation at room temperature. This radiation is not visible to the human eye, but exists nevertheless—for example, in the infrared domain. This can be nicely visualized with the typical infrared images we are accustomed to seeing from the military night-vision helmets that modern soldiers wear. Infrared images can be acquired with special imaging video or CCD cameras which are sensitive to wavelengths other than those that a human eye perceives (in Fig. 9 the head of a boy in perfect darkness). It becomes clear that our bodies emit EM waves as lamps do. If our eyes were sensitive to infrared radiation (like some species of snakes, fish or mosquitoes are), we could easily see how, even in perfect visible darkness, the objects around us would glow.

More generally, we can say that every object that has a non-zero absolute temperature emits some amount of radiation. Only in the very extreme case in which we froze it to the absolute zero temperature of 0 Kelvin, to a temperature of about -273,15° Celsius (a purely ideal case that, as we shall see later, is forbidden by the laws of QM), would it no longer emit any radiation. However, apart from that extreme case, everything emits some sort of heat in the form of EM radiation on all wavelengths.

It becomes reasonable, then, to assume that there must be some relation between the temperature T of a body, the wavelength λ of the EM wave it emits, and the intensity I of that radiation.

To discover what this relation must be, and what principles and phenomena stand behind the production of this radiation at the microscopic scale and at the molecular and atomic scale, physicists reasoned in terms of a so-called 'blackbody' (whole word, sometimes written also as 'black-body') model as follows.

Consider an empty body with a tiny hole in it—that is, a cavity where we inject a beam of EM waves through the tiny aperture. Moreover, consider the walls inside this cavity as being quite good but not perfect reflectors, so that light rays, once in the cavity, bounce back and forth many times, distributing themselves homogeneously and yet having the ability to be absorbed without finding their way back out through the small hole.

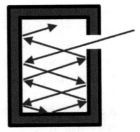

Fig. 10 The blackbody cavity.

This implies that all the radiation remains trapped inside the cavity and finally gets absorbed by the material which makes up the containing walls. In other words, it is a perfect black body because it absorbs all the radiation but does not re-emit it. In physics, a 'blackbody' is, by definition, a body that absorbs all the radiation (technically, one says it has a unitary absorption coefficient) at every wavelength and at every temperature, independently of the substance from which it is made or its form and structure. This is, of course, only an idealized model, as no real material object is a perfect absorber, but it is a good approximation for most cases of interest.

The next step is to consider the cavity in stationary conditions and in thermodynamic equilibrium with the environment, which means it must not change its temperature T or other physical parameters in time. If this condition is met, we can assume that the thermal EM radiation trapped inside the cavity will quickly distribute itself homogeneously, making the enclosure's temperature uniform throughout its structure. The system can, therefore, be characterized by its energy density (that is, the amount of energy per unit volume), in dependence on its temperature and wavelength, which do not depend on the shape, material or form and structure of the body. The energy lost through the tiny hole can be considered negligible, or eventually can be replaced, keeping the internal radiation energy density constant by injecting the same amount of energy (per time unit) as the cavity emits. The tiny aperture's emission can also be used to look into the cavity without disturbing the system, revealing to us the wavelength and intensity of the radiation it contains.

At this point, it becomes clear why every material object in thermal equilibrium with the environment can, despite a somewhat misleading terminology, be considered a 'blackbody'. In such a context, this should not be confused with the color or brightness of an object. Also, the Sun can be considered a 'blackbody' in the sense that physicists mean, as it is an object which maintains its temperature over a long period of time and the energy which it emits into outer space is replaced by the energy it produces in its inner core.

Physicists of the 19th century considered the blackbody radiation cavity to be a model that was supposed to appropriately describe the EM energy content of any physical body with a finite temperature. The step from a qualitative to a more quantitative approach which could describe, with a formula, the intensity of the EM radiation that a body emitted in function of its temperature and wavelength (or its color, in the visible domain), considered the energy density of radiation per unit frequency interval and unit volume. This radiation, when in a stationary state, was modeled with what physicists call 'stationary' or 'standing waves'. These behave like EM waves in a box, reflected back and forth but not interfering with each other destructively, building oscillating but stable wave structures. Standing waves are precisely all those that have only those wavelengths matching the cavity walls with a zero amplitude, as shown in Fig. 11. As an analogy, think about swinging a rod fixed at a wall. Eventually, try this experiment by yourself. Fix a rod on a wall and, at the other extreme, swing it up and down, producing a mechanical perturbation that propagates along the rod and will be reflected from the wall. The reflected wave will interfere with the incoming one and interference phenomena will occur. What you will easily discover is that for only some precise oscillation frequencies will you be able to create a stable oscillation pattern.

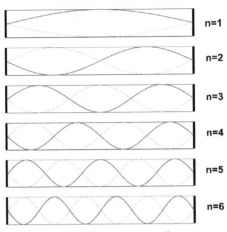

Fig. 11 Standing waves in a box.

Therefore, standing waves can exist only at specific wavelengths, called 'modes'. Only at these mode frequencies do they interfere with themselves constructively. For other frequencies, one observes disordered oscillations. So, not every kind of EM wave can be contained in a cavity in stationary mode—only those which match the right frequencies. Simply by looking at Fig. 11, one can see that the second mode has half the wavelength (twice the frequency) of the first mode. The third mode has one-third the wavelength (three times the frequency) of the first mode, and so on. Note that this also means there is a mode of minimum frequency, the first mode, below which no other modes with a lower frequency can exist. However, there is no upper bound. It is not too difficult to see that even in these conditions, an infinite number of such modes can fit between the walls. In Fig. 11, the first six modes are shown, but, in principle, there is no upper limit. The mode number n can become virtually as large as we wish. If we approach an infinitely small wavelength (an infinitely high frequency), we can accommodate an infinite number of modes.

The problem with this model is that in classical mechanics (CM), there is no reason to assume that one EM mode should be considered more privileged than the others. This means that, if we wait long enough (that is, if we wait until the system is in thermodynamic equilibrium), the energy will be exchanged and distributed among all the modes equally until they finally attain the same average energy. This describes the classical 'equipartition theorem'.

These modes also interact with the particles of the walls of the enclosure or with those within its volume. The particles making up the container can be seen as many little springs which oscillate with a different frequency, so-called 'harmonic oscillators', and which absorb and reemit the modes' energy. In fact, it is through the repeated absorption and emission from and back to the modes that this energy exchange among the modes itself can occur. Then, it seems reasonable to assume that this equilibrium mean energy of the modes stems from the amount of the body's thermal energy, which, in turn, is determined by its temperature. In this picture, each mode can, in principle, oscillate with any conceivable energy. Therefore, as tiny as this average oscillation energy might be, if the cavity is supposed to contain an infinite number of modes, because there is no upper limit to the frequency a mode can acquire, each with the same small but non-zero energy, we are forced to conclude that the cavity as a whole contains an infinite energy.

So, clearly, something must have gone wrong here. Yet even though this result seemed immediately implausible, nobody knew for sure how to tackle it.

As long as Planck didn't find the way out of this dilemma, the best model of the time for the blackbody radiation was that derived by Lord Rayleigh in 1900. This model is known as the Rayleigh-Jeans law. It clearly displays what is called the 'ultraviolet catastrophe': the energy density of the radiation of a blackbody at a temperature T diverges at high frequencies. It was called 'ultraviolet' because ultraviolet light makes up part of the upper-frequency spectrum of light, where the energy tends to diverge.

This depressing state of affairs lasted until 1900, when Max Planck came to the rescue. (This was a revolutionary period in the history of physics

Fig. 12 Max Planck (1858-1947).

because, five years later, Einstein published his work on special relativity.) The German physicist, who was then working on his dissertation thesis, introduced two fundamental and revolutionary hypothesis into the above description of the blackbody model.

First, he proposed replacing the arbitrary energy value of each EM mode (or harmonic oscillator) with integer multiples of the first mode energy. As we have seen, the frequency of the n-th mode standing wave is n times the frequency v of the first mode. Planck also posited a direct linear relationship exists between this n-th frequency and the n-th energy state of the n-th mode, expressed as:

$$E_n = n \cdot hv , \quad \text{Eq. 2}$$

with h being the Planck constant, and whose value is: $h = 6.626 \times 10^{-34}$J/Hz ([J] is the energy unit in Joule and, as we already saw, [Hz] is the frequency unit; see appendix A II for the fundamental physical units and constants). It is a value which is not predicted by any theory but is simply provided by Nature and must be measured as the mass of an electron or proton or the gravitational constant which determines the gravitational field of a given mass, etc. Let us first focus on the meaning of this little and simple relation which is extremely important in QP.

Eq. 2 is called the '*Planck-Einstein relation*'. (Einstein will give it a deeper meaning by explaining the photoelectric effect; see the coming section. We will simply call it 'Planck's equation', not to be confused with his radiation law equation.) It is a very simple equation but is also one of the most famous and revolutionary equations of 20th-century physics. It tells us that the EM field in the blackbody cavity must be composed of many modes which take energy values such as hv, 2hv, 3hv, etc., but cannot have any value in between, such as 1.5· hv or 2.7· v, etc. The EM modes in the cavity must appear in energy steps of hv (with v being the frequency of the first mode and which is given only by the geometrical structure of the cavity). Any value in between is not allowed to exist. Therefore, despite being the cavity filled with an astronomical number of modes, none of it can have an energy less than a minimum energy hv, which is called the '*ground state energy*'. This is a typical aspect of QT: Energy appears in '*quanta*' and a state of zero energy does not exist. Every quantum system—such as, for example, an atom—has a non-zero ground state energy, as we will see in more detail later.

Modern textbooks associate the n modes in the cavity with n particles of EM energy hv, called '*photons*'. Historically, however, as we shall see, this concept and the word 'photon' was introduced only later by N. Lewis and adopted by Einstein in describing the photoelectric effect. Actually, there is no need to introduce light particles at this stage.

The idea of replacing continuous with discrete energy modes was not entirely a novel one. It was already well-known that atoms absorb and emit light only with specific wavelengths, and not throughout a continuous spectrum. The spectral analysis of elements dates back to 1861, when Gustav Robert Kirchhoff (1824-1887) and Robert Wilhelm Bunsen (1811-1899) realized how every element emits and absorbs light at only very specific wavelengths. Each element has its own spectrum, which is like a characteristic 'optical fingerprint'. More generally, every element or substance, every type of atom or molecule, resonates only at certain frequencies which can be visualized with their spectral lines. For example, in Fig. 13 you can see the emission spectrum of hydrogen and mercury atoms.

When the hydrogen atom is excited (say, by an electric discharge or by heat), it re-emits the absorbed energy only at certain colors of the visible spectrum, each shown with one line. There is no continuity in the emission.

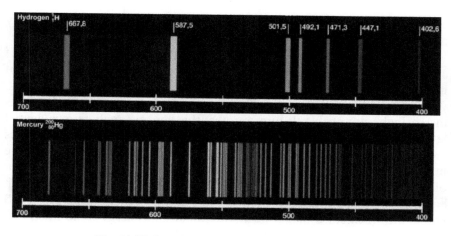

Fig. 13 Hydrogen's and mercury's spectral emission lines (wavelength units: nm=10^{-9} m).

The same thing happens with mercury, which, however, shows to have several more emission spectra lines than hydrogen does. This indicates that the mercury atom must be much more complex than the hydrogen atom.

Therefore, the analogy between these discrete spectral lines and the energy states of atoms with standing waves which exist only for specific frequencies and wavelengths was not too farfetched. However, a clear understanding of how these spectra come into being had to wait at least another 70 years after Kirchhoff's and Bunsen's observations.

The second hypothesis Planck introduced was to consider not all the modes as being equally probable. This wasn't a new idea either, as it had already been considered by the Austrian physicist Ludwig Boltzmann, who is well-known for his contribution to the field of statistical physics. According to Boltzmann, for a blackbody in thermal equilibrium (which, again, means at a constant temperature), the probability of finding a mode with a certain energy will decline as the amount of energy increases. According to this hypothesis (which Boltzmann described in a mathematically precise manner), there are many more modes with lower energy than those with higher energy. The energy of the cavity is, therefore, no longer distributed equally on all modes but, rather, according to a law which is called the *'Boltzmann distribution law'*.

However, the decisive difference between the idea of Planck and that of Boltzmann was that the former introduced the hypothesis which replaced the arbitrary continuous energy values E with the above-described quantized ones E_n. The probability of finding modes with energy E_{n+1} is smaller than that of finding those with energy E_n. Also in this model, for an increasing mode number n, the energy of the mode increases too, though the probability of finding such modes also becomes increasingly smaller.

At this point, it becomes clear why there could be no ultraviolet catastrophe. After a certain point, the probability of finding modes of increasingly high energy becomes so small that for all practical purposes, we can consider it to be zero. A mode with increasing n becomes increasingly improbable. It turns out that this happens with an exponential law, that is, the probability of finding modes with higher energy decreases increasingly fast. This also means that if we add up all the energies of the infinite number of modes, the sum will nevertheless reveal a finite amount of energy. In mathematics, one calls this a convergent series. To put it another way, an infinitely high energy mode has an infinitely small probability of existing; that is, no mode with infinite energy could exist, and, therefore, the total energy in the cavity stays finite. Nothing explodes in an infinite blast of energy and the Universe is safe!

Moreover, Planck's great achievement was not only avoiding catastrophic infinities in calculations but also furnishing a precise formula which correctly describes the blackbody radiation we observe in Nature. He again did all the calculations that Boltzmann had already done, but with the decisive modification that replaced the continuous energy function with the quantized one. By doing so, he obtained an expression which nicely describes the energy density of the EM radiation spectrum of a blackbody with temperature T at any given wavelength λ (or frequency ν): Planck's formula for the blackbody radiation law. (For those who would like to see what it looks like, see appendix A III.)

Fig. 14 shows a graph where, on the horizontal axis, the abscissa, the wavelength of a body, is given, while on the vertical axis, the ordinate represents the intensity (in arbitrary units) of the EM radiation it emits.

Each curve on the graph represents a body with a specific temperature. The rightmost curve represents a body for 5000 degrees Kelvin and is calculated with CP. As one can see, it quickly diverges for smaller wavelengths, leading to the ultraviolet catastrophe. On the other hand, the remaining 5000 K curve represents the radiation of the body with the same temperature but calculated by taking into account the fact that energy is absorbed in quanta and into oscillation modes with energy states of different probabilities, according to Planck's idea.

You can see that, apart from cancelling the divergence, it has a maximum peak at about 0.6 μm ('micrometer'), which corresponds with the yellow color. In fact, the surface of our Sun has a temperature of about 5200° Celsius and, as is well-known, appears to have a yellow color. A couple of other curves for the radiation of bodies at 4000 K and 3000 K are shown.

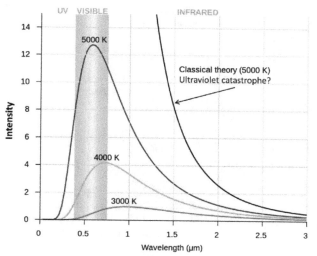

Fig. 14 The blackbody spectrum.

Note how the colder the body, the more shifted towards large wavelengths is its maximum. Therefore, hot bodies tend to radiate in higher frequency domains (more blue or ultraviolet), whereas cold bodies tend towards the visible or infrared light. If a body becomes white, this means that it has become so hot that it radiates strongly in any or several wavelengths. This is because, as is well-known, when all colors are mixed together, the resultant color appears as white—which, by the way, also makes it clear that, strictly speaking, white isn't physically a color, just as black isn't.

Planck's theory solved the problem of the ultraviolet catastrophe and was also confirmed experimentally, measuring the radiation emitted by different bodies at different temperatures.

The fact that Planck's spectral curves do not depend on the type of material but only on its temperature gave astronomers a great tool for measuring the temperature of distant objects, like stars. There is no need to physically reach, with a thermometer, a star which is light years away. We just have to measure its spectral curve and we will know its temperature simply by fitting it with Planck's equation.

For instance, Fig. 15 shows the spectrum of solar radiation. It does not appear as smooth as that of Fig. 14. However, this is only because it is measured through the earth's atmosphere. The solar spectrum is absorbed by the molecules that make up the atmosphere itself and produce some 'holes' in the spectrum.

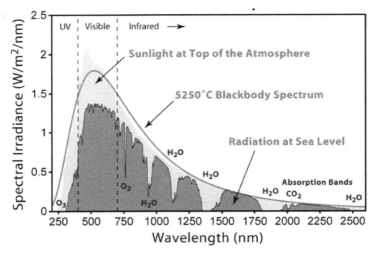

Fig. 15 The solar radiation spectrum.

While the theoretical blackbody radiation fits well with the measured curve of the sun at the top of the atmosphere, at sea level, there are several *'absorption bands'* corresponding to several kinds of molecules, typically water or carbon dioxide molecules. With this kind of observation and experimental analysis, it became possible to determine not only the temperature but also the chemical composition of stars or other distant objects—a formerly impossible feat.

As a historical side note, it might be interesting to point out how unexpected and surprising developments in science can be. Before the discovery of Planck's blackbody law and spectroscopy, the simple act of thinking about the possibility of measuring the chemical composition and temperature of stars at cosmic distances was considered absolutely impossible and metaphysical. In 1835, the French philosopher Auguste Comte had no doubts that: *"On the subject of stars, all investigations which are not ultimately reducible to simple visual observations are ... necessarily denied to us. While we can conceive of the possibility of determining their shapes, their sizes, and their motions, we shall never be able by any means to study their chemical composition or their mineralogical structure ... Our knowledge concerning their gaseous envelopes is necessarily limited to their existence, size ... and refractive power, we shall not at all be able to determine their chemical composition or even their density ... I regard any notion concerning the true mean temperature of the various stars as forever denied to us".* [2]

The blackbody radiation was also confirmed much later, at about the end of the 1960s, on a cosmic scale. In fact, we can conceive of the whole Universe as a single body—or, if you prefer, a 'cavity'—which contains a sea of EM radiation and matter, like every other object. According to the Big Bang theory, when the Universe came into being by a huge explosion about 13.7 billion years ago, it was extremely hot. It must have had extremely high temperatures but then expanded and cooled down. Yet as a whole, it can still be considered a body with some temperature which must contain radiation, just like, in the case of the cavity, the so-called *'cosmic background radiation'*, with which we can associate a temperature. Does the spectrum of the cosmic background radiation also follow the bell-shaped Planck spectrum? The answer came about by measuring radio waves with radio telescopes from the ground and was later confirmed with much more precise measurements on board satellites free from atmospheric absorption. The result is visible in Fig. 16, the *cosmic microwave background* (CMB) spectrum measured by the COBE satellite.

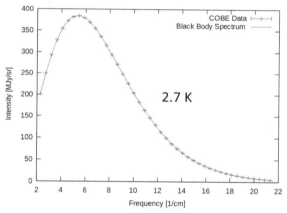

Fig. 16 Cosmic microwave background spectrum from COBE.

You can see how there is an almost perfect match between the measured flux coming from the Universe and that calculated for a body with 2.7 K temperature (units in frequency are wavelengths/cm and intensity in mega Jansky/steradian, that is, 10^{-20} W/solid angle unit). So, the Universe is a very cold place, but on average still not at absolute zero.

Just for the curious who want to know more, in Fig. 17 you see the map of the overall structure of the Universe inferred from cosmic background radiation—more precisely, from its fluctuations from the ideal Planck spectrum. These ripples and tiny inhomogeneities represent the small departures from perfect homogeneity of matter and energy distributions. This latter aspect will be of great relevance when we discuss quantum cosmology and cosmic inflation in the second volume.

Fig. 17 CMB inhomogeneities (Wilkinson Microwave Anisotropy Probe-WMAP).

These observations confirmed Planck's theory, which can be considered one of the greatest triumphs of the physics of the last century. Note also how a precise theoretical prediction was followed by solid observational data which confirmed it. This is something which, as we shall see later, has become increasingly less obvious in recent times. It also makes clear that Boltzmann was incredibly close to the solution. Had he replaced the continuous energy function with a discrete valued one in his own mathematical theory, all Planck's merits and fame would have been Boltzmann's instead. This shows how sometimes very simple but subtle intuitions can change the history of science, leading to great paradigm shifts. Boltzmann was also a fervent supporter of atomic theory, something which was considered a controversial matter at the time. Had he not committed suicide, he probably would have lived long enough to see that history was on his side.

So, the most important message of this first section, one that we will have to keep in mind throughout, is that we must surrender to the idea and fact that, according to QM, energy (and, we will see later, other physical quantities) seems to come up in the form of discretized chunks and packets, or *'quanta'*. Energy is absorbed or emitted only in discrete quantum amounts by matter. Energy is a 'quantized' physical quantity. This fact became the conceptual foundation on which all modern QP is based, and this will follow us in the coming chapters, too. That is, after all, where the names 'quantum mechanics', 'quantum physics' and 'quantum theory' come from. Quantum physics was born!

2. More evidence for particles: the photoelectric effect

We have seen how the theory of blackbody radiation fits the experimental evidence only if we assume that energy is always absorbed and emitted 'packet wise', in the form of discrete quanta. However, blackbody radiation was only the first hint of this kind. Many other experiments showed how Nature exchanges energy only in the form of quanta. These experiments told us the same story in different ways, but always with the same underlying plot, according to which energy—and, therefore, also light—can be conceived of as made of what we imagine to be particles.

One of the most important pieces of evidence of this type came from the so-called *'photoelectric effect'*, which conventional physics considers to be proof of the existence of the particle of light, the *'photon'* (word coined by the American physicists Gilbert N. Lewis). The photoelectric effect had already been discovered in 1887 by the German physicist H.R. Hertz.

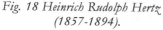

Fig. 18 Heinrich Rudolph Hertz (1857-1894).   *Albert Einstein (1879-1955).*

He observed that sending light of sufficiently high frequency onto a metal plate leads to the emission of charged particles from the metal surface. These turned out to be electrons. Hertz, however, was not able to explain the true meaning and physical implications of the phenomenon. The theoretical understanding and explanation came later, in 1905, developed by Einstein. To understand what this is all about, we will outline only a brief sketch here, without going much into the technical details. The goal is simply to give you an intuitive idea.

Imagine having a metal plate onto which you shine a beam of light. Suppose the light is in the far infrared spectrum, much below the red-colored light, which means we are using only a relatively low-frequency (long wavelength) EM wave. This low-frequency light beam would have no effect on the metal plate (except, of course, that the plate would heat up or eventually melt); no emission of charged particles, the electrons, would be observed as long as the light was under a specific threshold frequency (or, equivalently, above a specific wavelength). This occurs regardless of the beam's intensity; it will not happen even for strong light intensities. If, however, the light goes beyond a threshold frequency v_0 (below a threshold wavelength $\lambda_0 = \frac{c}{v_0}$), suddenly a flux of electrons from the surface of the metal can be detected. And, again, this occurs regardless of the beam's intensity; it also happens for low-intensity light.

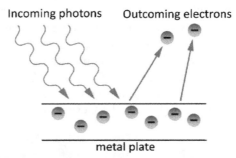

Fig. 19 The photoelectric effect (1887/1905).

What this shows is that light waves can extract the electrons, the so-called '*photoelectrons*', from the atomic metal lattice. The fact that this occurs only for certain metal plates makes it clear that the photoelectric effect does not extract electrons directly from the atoms. Photoelectrons are those electrons which were trapped inside the atomic lattice and are nevertheless free to move inside it when an electric field is applied. This property makes good conductors out of certain metals and allows us to transport electric energy through metal cables.

Therefore, this extraction process from the metal lattice occurs only when the electrons acquire a specific energy at a minimum extraction frequency v_0 (which, of course, is related to a minimum extraction wavelength, as $\lambda_0 = \frac{c}{v_0}$). Fig. 20 illustrates this state of affairs; if we trace along the horizontal axis the frequency of the incident light and along the vertical axis the kinetic energy, E_{kin}, of the outgoing photoelectron, we can see how only if the wavelength becomes small enough (or, equivalently, only if the frequency becomes high enough, that is, larger than v_0), the electrons will start being 'kicked' out from the surface and appear with some kinetic energy larger than zero. This implies that part of the light energy that is incident on the metal surface must be employed to extract the electrons first from the metal lattice. This minimum energy is called the 'work function ϕ'. Every metal has a different work function, which means that every metal has a different threshold frequency. However, the effect remains qualitatively the same.

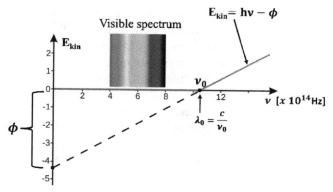

Fig. 20 Kinetic energy of the extracted photoelectrons.

Once the electrons are extracted from the plate, what remains of the energy of what was once light will go into the electrons' kinetic energy. So, in general, it turns out that the observed electrons' kinetic energy E_{kin} must be the difference between the energy contained in the incident light photons $h\nu$ and the amount of work ϕ necessary to extract them from the metal.

In brief:

$$E_{kin} = h\nu - \phi.$$

Note that this is not theoretical speculation, but an exclusively experimental fact: measuring for each light frequency with which one shines the metal plate, and obtaining from that the kinetic energy of the electrons for that frequency, one can plot it into a graph and obtain a strictly linear dependency between the light's incident energy and the emerging electrons' kinetic energy, like that depicted in Fig. 20. While the threshold wavelength λ_0 (or threshold frequency ν_0) is determined by the type of metal one uses, the slope of the straight line is always the same. Every metal and kind of substance one uses has a different ϕ—that is, the straight line function is only shifted parallelly upward or downward, but does not alter its form or steepness. The slope of the kinetic energy linear function is a universal constant, namely, Planck's constant. This is, in fact, how Planck's constant could be measured precisely.

Moreover, increasing the intensity, which results in an increase in the number of incident photons, does not change the result. An increase in the light intensity, that is, the photon's flux, does not at all alter the above graph. It simply changes the number of observed photoelectrons (if above the threshold), and not their kinetic energy. A higher light intensity does not lead to more energetic ejected electrons; it only determines its number.

This is somewhat unusual for a classical understanding of where we imagine light as a wave. The question at this point is: Why does light extract electrons from the metal only above a threshold frequency (say, in the spectral domain of ultraviolet light or x-rays) but not below it? Why can we not bombard the metal surface with a higher-*intensity* light of the same wavelength (just imagine many waves with large amplitudes) to obtain the same result? How should we interpret this result?

The generally accepted explanation, according to Einstein, is that, as we have seen with the blackbody radiation, we must conceive of light as made of single particles—that is, photons carrying a specific amount of energy and impulse, and which are absorbed by the electrons one by one. However, this happens only if the single photon has sufficient energy to extract the electron from its crystal lattice in the metal. The electron will not absorb two or more photons to overcome the energy barrier of the lattice in which it is trapped. It must wait for the photon that has all the necessary energy that will allow it to be removed from the metal lattice. And because it absorbs only one photon at a time, all the electrons emerging from the plate will never show up with energies higher than that of the single absorbed photon, even though a larger number of photons (higher intensity of light) can extract a larger number of electrons.

In 1905, Einstein published his theory of the photoelectric effect alongside his historic paper on SR. For this explanation of the photoelectric effect and his 'corpuscular' (little particles) interpretation of light, Einstein received his Nobel Prize—not for the theory of relativity, as is sometimes wrongly believed.

3. Bohr's atomic model

At this point, physicists had a sufficiently broad experimental and conceptual understanding of the physics of atoms to go a step further. The fact that matter emits and absorbs EM energy in a discrete manner had become, at that time, an established and accepted fact. What was missing, however, was a structural model which could, at least partially, explain this. Ernest Rutherford (1871-1937), a physicist from New Zealand, discovered in 1911, by performing scattering experiments with radioactive particles, that atoms are almost empty objects, as over 99.9% of the mass of an atom is concentrated inside a tiny positively charged nucleus. Putting things together, a couple of years later, Niels Bohr, a Danish physicist, conceived of the atom as a kind of tiny solar system, with the negative electrons circling the positively charged nucleus made of protons. This is the model most of you may have encountered in school. What characterizes Bohr's model is that it retains the idea of electrons flying along very well-defined and precise orbits. Every orbit can be occupied by a certain number of electrons. Moreover, every orbit corresponds to some energy level of the electrons in the atom. There were no intermediate orbits, only orbits corresponding to specific energy levels, which can be labeled with progressive integers, n = (1, 2, 3, ...).

These numbers were called '*principal quantum numbers*' because these define the possible discrete energy states of the atom.

This was the first atomic model that could explain why atoms absorb and emit light in discrete quantities. If the orbits are fixed, the energy levels of electrons in those orbits around the nucleus must also be discrete. The energy absorption and re-emission is conceptually visualized with the transition of an electron from one orbit to another. If an electron absorbs a photon coming from the environment, it transits towards a higher orbit, but then will quickly re-emit the photon back by returning to the lower-energy-level orbit. Re-emission occurs because every system will always tend to return to the lowest energy level possible. Because an empty place is left by the electron, it will re-occupy it back by emission. It is not necessary that the transition be between neighboring orbits; it could also occur between orbits with very different quantum numbers. The difference in energy of the two orbits will be simply that energy which the electron absorbs and emits. This energy difference will also be that which determines the frequency (or wavelength) of the absorbed and emitted photon according to Planck's equation of Eq. 2.

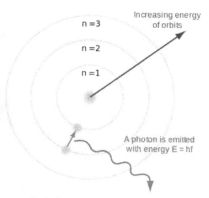

Fig. 21 Niels Bohr (1885-1964).   *Bohr's atomic model of 1913.*

This model could indeed explain the atomic spectra of the hydrogen atom measured experimentally (see Fig. 13). For every electron transition from one orbit to the other, in each of its possible combinations, one can calculate the energy difference and, from there, the energy and wavelength of the absorbed or emitted photon. Fig. 22 displays a qualitative picture of some of these possibilities.

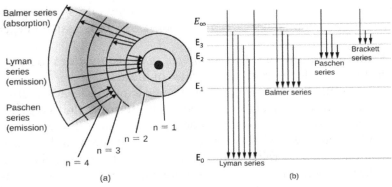

Fig. 22 The energy levels of the hydrogen atom.

For example, the transition to the ground state (quantum number n=1) corresponding to the energy E_0 from all the above discrete electron orbits (n=2, n=3, etc.) of energy E_1, E_2... etc., emits photons of energy E_1-E_0, E_2-E_0, etc. This is the so-called 'Lyman series' of hydrogen's spectral lines. Many other possible series exist, each labeled with the name of its discoverer.

As an additional historical side note, it might be worth mentioning that Bohr's atomic model was refined and extended by Arnold Sommerfeld, a German physicist, who also considered elliptical electron orbits, just as we know it from planetary orbits. This is sociologically interesting because it shows how we, as humans, are instinctively driven first by the desire to transfer our understanding of the macrocosm onto the microcosm. In fact, this attempt was not without success. Sommerfeld introduced a second quantum number, the 'orbital' (or 'azimuthal') quantum number l, which for each principal quantum number n specifies n allowed orbital angular momenta of the electron around the atom's nucleus (l=0, 1, 2,...n-1, and designated as s, p, d, f, g,...), and which characterizes the ratio between the semi-major and semi-minor axis—that is, the elongation of the corresponding electron orbit. And, what a coincidence, this more general Bohr-Sommerfeld model was able to accurately explain a number of more complex atomic spectra. It provided a partial, at least an approximated, explanation of the splitting of some spectral lines that were observed due to the presence of strong external electric fields (the 'Stark effect') or magnetic fields (the 'Zeeman effect').

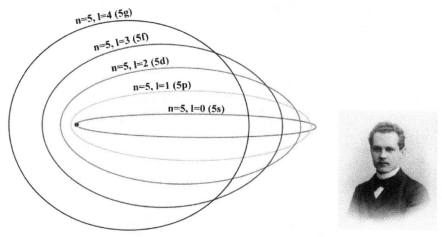

Fig. 23 The Bohr-Sommerfeld hydrogen model (for n=5).   Arnold Sommerfeld (1868-1951)

This was quite an exciting success at the time, as it seemed to indicate that physicists were after something that could lead them in the right direction. However, many inconsistencies soon became apparent. The Bohr-Sommerfeld model was able to make these predictions with only limited precision and, at any rate, only for the hydrogen spectra. Such a simple theory cannot explain all the other elements of the periodic table, nor can it explain the molecular energy levels and structure.

It was an attempt to explain atomic behavior and energy states. It is, indeed, an interesting historical remnant of the intellectual journey of humankind, and it made perfect sense when viewed in that historical context. However, it is no longer tenable today. Unfortunately, it is still taught in schools as if it were the final sentence and summation of atomic physics. And it was a good example which shows that some theories can make correct new predictions and yet still be wrong. This is something that many scientists seem to forget but that we should always keep in mind when working with modern theories.

QM will explain things much more accurately, as we shall see later. It was only upon the advent of QP and its application to the atomic structure that science could correctly predict atomic spectra. The atomic model that arises from QM will be completely different from that of Bohr, which could be considered only a temporary sketch, as Bohr himself had to admit.

4. Even more evidence for particles: the Franck-Hertz experiment, Compton scattering and pair production

Another nice and clear-cut piece of experimental evidence that atoms must have discrete energy levels came only a year later, from the so-called 'Franck-Hertz experiment'. G. L. Hertz was the nephew of H. R. Hertz, the German physicist mentioned as the discoverer of the photoelectric effect. While H. R. Hertz, among other things, proved the existence of EM waves, G. L. Hertz and James Franck were awarded the 1925 Nobel Prize in physics *"for their discovery of the laws governing the impact of an electron upon an atom"* [3], using the experiment we are going to describe.

Fig. 24 Gustav Ludwig Hertz (1887-1975).

James Franck (1882-1964).

The experimental setup (see Fig. 25) consists of Hg gas atoms (Hg is the symbol for the element mercury) inside a low-pressure bulb. An electric cathode—that is, something like the heated filament of a light bulb—emits electrons.

Therefore, this part of the device emits not only light but also negatively charged particles. An electric field is applied between the electron emitting cathode and a positively poled grating, with a battery or some other electric source, which builds up an electric potential. Due to their negative charge, this difference in the electric potential field leads to the electrons' acceleration and conveys to them some kinetic energy (as you might recall from school, charges with the same polarity repel each other whereas those with opposite polarity attract each other). When the electrons reach the grating, most of them will fly through because the mesh of the grating is kept sufficiently wide to allow for that. This first part functioned as a little electron accelerator. Then, between the grating and a collecting plate on the right side, another field is applied.

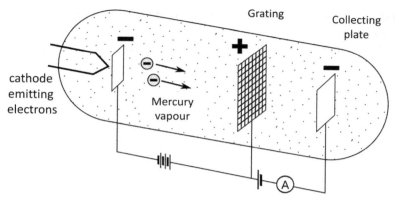

Fig. 25 The Franck-Hertz experiment setup of 1914.

However, in this second part of their journey, they will experience an inversely polarized electric field as, after passing the grating, they will be repelled because they will begin to 'feel' the negatively charged collecting plate. This means that, through use of an Amperemeter, a device which measures electric current (the number of electrons), one can measure the flux of electrons which flow between the grating and the collecting plate. While the electrons' initial energy is proportional to the applied electric field intensity (the voltage) between the emitting cathode and the grating, in this second part, they are decelerated due to the opposite polarity. The measurement of the current, therefore, allows one to determine the number of electrons that make it through to the collecting plate, which provides information about how their energy is affected by the atoms while flying through the gas in this second part of the bulb. This is done by varying that field, step by step, for several voltages.

Franck and Hertz's insight was that, while flying through the gas of atoms, several electrons must sooner or later hit one or more atoms and be scattered either elastically or inelastically. Elastic scattering means that when objects hit a target, they change course but maintain the same kinetic energy, while inelastic scattering implies that they lose part or all of their kinetic energy in the collision process. It follows that there must be a measurable difference between the energy of the injected electrons reaching the grating and the energy of those which flew through the gas, hitting the collecting plate. This difference is made clear to the observer by measuring the current between the grating and the collecting plate. This energy gap tells us something about the energy absorbed by the atoms in the gas.

Therefore, if atoms absorb energy only in the form of quanta, this implies that, while we slowly increase the kinetic energy of the injected electrons, we should be able to observe when and to what degree the electrons' energy is absorbed by the gas of Hg atoms.

This is, indeed, what happens, as can be observed in the graph of Fig. 26.

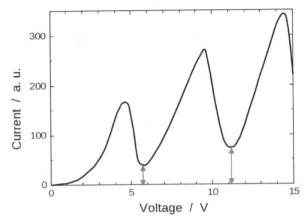

Fig. 26 Energy level of the mercury atoms in the Franck-Hertz experiment.

While the injected electrons' kinetic energy is increased steadily by application of an electric potential from 0 to about 15 V between the cathode and the grating (horizontal axis), the current of the electrons measured at the collecting plate (vertical axis) increases accordingly, though not in a linear fashion. We observe that the electrons do not have a final kinetic energy which increases proportionally to the electrons' input energy, according to what one would expect for an elastic scattering between classical objects (think, for example, of billiard balls). What we see instead is that at first (between 0 and 4 V), the relation between the input and output energy is approximately linear, which means that the electrons are scattered through the gas elastically; they do not lose considerable kinetic energy.

At about 4.5 V, the first bump appears. Between 4 and 5 V, the output energy of the electrons decreases steadily, despite their increasing initial energy. This signals an inelastic scattering: Some of the electrons' initial energy must have suddenly been absorbed in collisions with the Hg atoms. However, this does not happen before a certain kinetic energy threshold of the electrons hitting the Hg atoms. At about 5.8 V, almost all the kinetic energy is lost and goes into the internal excitation of the atoms. There is, however, a remaining minimum energy gap which is shown in the figure with the vertical arrow. The difference between the first peak and the first minimum is the maximum amount of kinetic energy the atoms are able to absorb from the electrons. Therefore, it furnishes the first excited energy level of the Hg atom. Then, after about 6-9 V, the energy begins to increase again, meaning that the atoms absorb only that aforementioned discrete amount of the electrons' energy, but not more than that. The remaining energy goes again into elastic scattering. All this repeats regularly at about 9-10 V and about 14 V.

The existence of these 'bumps' at different input energies means that atoms must have several different but discrete energy levels. Franck-Hertz's was the first direct experimental proof confirming Planck's idea that matter absorbs energy in discrete quanta. Moreover, this validated the discrete spectral lines of light spectra, as did Bohr's idea of representing the atoms in the form of a model which resembles a tiny solar system—that is, with electrons moving only in specific orbits with their respective quantum numbers which represent different but discrete energy levels.

Not too many years later, other types of phenomena confirmed energy quantization in Nature. One of these, in 1923, was the so-called 'Compton scattering', documented by the American physicist A.H. Compton.

Compton also started from the assumption that EM waves could be considered a flux of light particles. If this were true, it would then be possible to calculate precisely the scattering process between a single photon and an electron, just like it is possible to describe the elastic collisions between two billiard balls, using the simple laws of energy and momentum conservation of CP.

Fig. 27 Arthur Holly Compton (1892-1962).

Now, by using these conservation laws, Compton wrote a concise and useful formula which relates the wavelength λ of an incoming high-energy gamma-ray photon (a photon with a wavelength sufficiently small to be comparable to the size of an electron) before the scattering with the electron, and the wavelength λ' of the scattered photon, according to a scattering angle θ (see Fig. 28). The wavelength λ' of the scattered photon must be larger than that of the incoming one, as it loses some of its energy. This is because the higher the energy of a photon, the smaller its wavelength will be. In fact, remember that the energy of a photon is given by Planck's equation $E = h\nu$, with ν the frequency. The wavelength λ of an EM wave is given by $\lambda = \frac{c}{\nu}$, with $c = 3x10^8 \frac{m}{s}$ (ca. 300.000 km/s), the speed of light in vacuum. So, the wavelength we associate with a photon must be inversely proportional to its energy as:

$$E = h\nu = h\frac{c}{\lambda}. \qquad \text{Eq. 3}$$

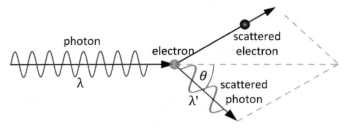

Fig. 28 The Compton scattering process.

At this point, the reader might be somewhat confused by the fact that we are talking about light particles, which are supposed to be point-like objects, and yet we continue to treat them (and even depict them graphically) as waves having a specific wavelength. Again, as we have already seen with the photoelectric effect, Nature seems to play with this ambiguity. It seems as if light, or any EM energy traveling throughout space, can be thought of as a wave and, because matter absorbs and emits it in the form of discrete energy quanta, also as a particle. We shall clarify the deeper meaning of this 'duality' in the coming sections. For now, please accept it on faith and follow the line of reasoning and phenomenology of these historical experiments, which became the foundation of all that will be explained later.

By putting all of this together, Compton was able to predict in 1923 a very strict relation between the difference in wavelength of the incoming and scattered photons and the scattering angle θ, which is the wavelength shift $\Delta\lambda$ radiation undergoes when it is scattered by matter. Compton's formula was as follows:

$$\Delta\lambda = \lambda' - \lambda = \lambda_0(1 - \cos\theta)$$

Here $\lambda_0 = \dfrac{h}{m_e c}$ is a constant and m_e is the mass of the scattering particle, in this case the electron. λ_0 is called the '*Compton wavelength*', which for the electron is about 2.426×10^{-12} m. Notice that if θ is zero, we have no difference, which means that the photon goes straight through without scattering and frequency change. Whereas for $\theta = 180°$, for backscattering, the wavelength increases by twice the Compton wavelength. The exact relation between the input and output wavelengths (or energies) of the particles and their respective scattering angle was verified experimentally, and it turned out that Compton's predictions came to fruition precisely. This is a great historic example of the triumph of theoretical physics confirmed by experimental verification—of how, in the history of science, we have found that sometimes first comes the math and then comes the verification in the lab (even though sometimes it goes the other way around).

Compton scattering is also an atomic scattering process. This suggests an explanation of how photons can ionize atoms. To ionize an atom means that a photon extracts an electron from an atom's outer orbit shell around its nucleus. A light particle with sufficiently high frequency can be absorbed by one of the electrons, so that the electron acquires a certain amount of kinetic energy and can eventually be removed by overcoming the atomic force potential that keeps it bound to the nucleus. A gamma-ray photon, however, can be so penetrating that it can ionize even inner electron shells and partially or completely transfer its EM energy to the electron, which acquires the gamma ray's momentum and kinetic energy and which for this reason is also called a 'photoelectron' (as in the photoelectric effect). Notice however that the photoelectric and the Compton effect are two very different physical processes. The former extracts the electrons from a metal lattice; the latter is a deep scattering process that can 'kick out' the electrons which are contained in the inner shells of an atom.

In Fig. 29 you can see an illustration of this process. A single high-energy photon can also be scattered by several atoms and will lose its energy with each repeated collision. A high-energy small-wavelength (high-frequency) photon can become, via multiple scatterings whereby it loses a bit of its energy in each collision, a low-energy large-wavelength (low-frequency) photon.

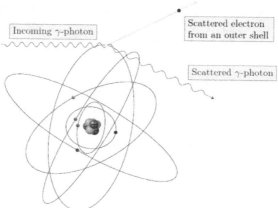

Fig. 29 Compton scattering of outer atomic shell electrons.

This is precisely what happens in the center of the Sun. In the Sun's core, which has a temperature of about 16 million degrees, there is a huge amount of gamma radiation. However, it takes several million years of scattering processes before the photons produced inside the Sun reach its surface, by which time they have lost most of their energy being 'downgraded' to the photons that we know as visible light. The light we observe today coming from the Sun's surface was once an intense radiation foam of gamma photons in its interior.

What makes Compton scattering so important and interesting for our considerations here is that all this seems to suggest again that we can reasonably think of EM waves no longer as waves at all, but rather in the context of light particles, or photons, kicking around other particles like tiny billiard balls.

If we were to stick to the classical idea that light is made of waves, we should observe a scattering of concentric waves by the electron (think of water waves scattered by an object). Compton's scattering instead indicates that the application of the conservation laws of CP, which determine the scattering between classical particles, holds also with photons. There is therefore no longer any reason to reject a 'particle picture' in QM in favor of the 'wave picture'.

Another effect which is worth mentioning is the 'pair creation' effect, which can occur as an alternative to Compton scattering. Pair creation is a physical phenomenon whereby a high-energy massless photon is converted into particles with a mass. Instead of the photon and electron being scattered, what occurs in this case is the absorption of the photon (by another particle or atom nucleus) and an immediate release of its energy in the form of material particles. The pair creation effect was discovered in 1933 by the British physicist P. Blackett. This phenomenon clearly shows that we must conceive of light not only as made of photons

Fig. 30 Patrick Blackett (1897-1974)

but also in the context that photons can transform into other material particles, like electrons. Pair creation, or even annihilation, which is the opposite process whereby particles with mass are converted into photons, is a striking example of Einstein's *mass-energy equivalence* which is described by his famous formula

$$E = mc^2 \qquad Eq.\ 4$$

where m is the *rest mass* of a particle, and which is called also the 'rest energy' (in relativity, the mass of a fast-moving object cannot be handled as the classical rest mass, but we won't go further into this here) and, again, c is the speed of light. The mass-energy equivalence tells us that every particle, atom, and any material object with a mass contains an energy in potential form which is given by Einstein's formula. The fact that it factors in the speed of light squared makes the amount of energy contained in matter huge. Only a gram of matter, if converted into kinetic energy, would cause an explosion like that of the bomb of Hiroshima.

In the case of a gamma photon pair creation (Fig. 31 left), two electrons can 'materialize' if a single photon interacts with matter, say a heavy nucleus, and converts into matter itself. One of these material outgoing particles will be the negatively charged electron e^-, while the other is a positively charged anti-electron e^+, also called a 'positron'. Anti-matter in general is made of the same particles with the same mass—that is, of electrons, protons, neutrons, etc.—but with the opposite electric charge and opposite spin. This process, however, can only happen if the energy of the incoming photon is sufficiently large to produce the two electrons according to Einstein's mass-energy equivalence formula. If the photon's energy is lower than twice the rest energy of two electrons, the process cannot occur and the photon will not be absorbed but instead eventually will be scattered.

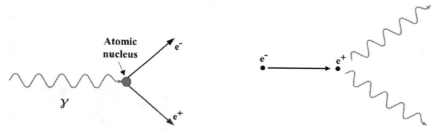

Fig. 31 Pair creation (left): a gamma-photon converts into an electron and anti-electron. Pair annihilation (right): an electron and anti-electron convert into two gamma photons.

The 'annihilation process' (Fig. 31 right) is the opposite process: Two massive particles (one ordinary and the other an anti-particle) interact with each other and are converted into gamma photons, which subsequently fly apart in opposite directions. For example, if an anti-electron comes into contact with ordinary matter, it annihilates to produce two gamma-photons with an energy given by Einstein's equation.

Fortunately, for some reason that is still not entirely clear, the Universe we live in is made almost completely of one type of particle; otherwise, we would have already 'evaporated away' in a foamy Universe where particles annihilate and create themselves continuously and stars, planets, and of course life, at least as we know it, couldn't have formed in the first place.

Thus far, we have listed several phenomena which indicate how energy is absorbed and emitted in a quantized manner and which therefore led to a corpuscular interpretation of light: Blackbody radiation, the photoelectric effect, the Franck-Hertz experiment, the Compton effect, and pair creation and annihilation can all be viewed in this context.

So, on one side we saw how light behaves as a wave; on the other side, EM radiation behaves as if it were made of particles. Which of the two points of view are correct? What is Nature trying to tell us here? This will be the main focus of the following sections.

5. Waves strike back: Bragg diffraction and the De Broglie hypothesis

We are now in a position to analyse the phenomenon of interference in real-world physical phenomena. We can begin with a classical example—what one observes in 'thin film interference'. This will open up the way for us to understand Bragg diffraction and the wave-particle duality.

In Fig. 32 you can see a couple of light rays with a certain angle of incidence reflected by a thin film surface. This thin film surface could be for example made of water, or the mixing of oil and water on the road after a rainstorm, or, as in this case here, that of a soap bubble.

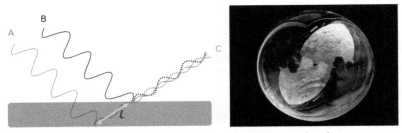

Fig. 32 Interference due to light reflection on thin surface layers. Example of the soup bubble.

The nice coloring of oil films or, as we enjoyed as children, of soap bubbles, is an interference phenomenon that can be explained as follows. These films have a very small but non-negligible thickness limited by an upper and lower layer. The reflection of light onto these layers can be visualized as a first beam, beam *A*, that goes unperturbed through the upper layer and will be reflected by the lower layer of the film, and a second beam *B* which is reflected by the upper layer. If we examine the situation carefully, it is easy to recognize that there is a difference in the travelled path between the two light rays. Even if both light rays are emitted at the same time from the same source and arrive with the same angle—that is, they are parallel to each other—when they reach the two thin film layers, they are nevertheless reflected at different depths and therefore acquire a path difference *l*, which implies a phase difference relative to each other (the little segment of length

l, in Fig. 32). The two beams then sum up and superimpose after the reflection, interfering according to their phase difference. This phase difference is determined only by the incident angle once we fix the thickness of the film and its refractive index (a dimensionless number which tells how much light is refracted and reflected or, equivalently, how fast it travels in a medium). This implies also that the outgoing light can interfere constructively, destructively, or somewhere in between, for every specific reflection angle. We will observe the resultant waves varying in intensity and wavelength along the different angular positions where we place our eyes. This is why we can observe the beautiful plurality of color patterns on the surface of a soap bubble.

The same thin film interference works not only with water layers in visible light but also for *atomic crystal lattice* layers at X-ray wavelengths. In the latter case we need very short EM wavelengths because the distances that separate the different atomic crystalline layers (also called '*crystallographic planes*') are very small, of the order of few atomic sizes. Obviously, at these wavelengths, we will no longer deal with color patterns since X-rays are well beyond the human-visible light spectrum, but the underlying principle remains the same.

In 1913, W. H. Bragg and his son W. L. Bragg, two British physicists, discovered how crystalline solids produce patterns of reflected X-rays that can be interpreted only as interference patterns, nowadays called '*Bragg diffraction* patterns'. They observed intense maxima and minima of reflected radiation, depending on the wavelength and angle of incidence—i.e., they could not be observed for reflected radiation of longer wavelengths.

Fig. 33 William Henry Bragg and William Lawrence Bragg
(1862-1942)   (1890-1971)

In fact, more precisely, at the atomic scale, we can no longer speak of classical reflections on a surface layer; we must conceive of the photons (or waves) as absorbed by the atom and re-emitted in all directions in the form of spherical waves. All these spherical wavefronts then sum up and form an interference pattern somewhere, for example on a screen.

Let us take a typical example of an atomic lattice: a salt crystal, which is nothing other than sodium chloride, a crystal made by sodium and chlorine atoms (here depicted with larger and smaller atoms, respectively).

The function of the thin film we considered in the example of the soap bubble with liquid layers can now be taken up by the atomic layers of the salt crystal lattice. Here again, two incident and parallel light beams will be reflected by two underlying crystallographic planes, as shown in Fig. 34.

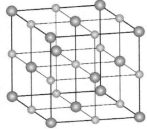

Fig. 34 A salt crystal and its atomic NaCl lattice structure.

There are essentially three physical variables that determine how the beams will be diffracted and will interfere—that is, how the outgoing waves will display their relative phase shift which gives rise to the interference pattern. First, all depends from the wavelength of the EM wave λ. Second, what determines the phase shift induced by the path difference is also the distance d which separates the atomic layers. This causes a path difference (the short wavy segment that connects the two atomic layers in Fig. 35), and consequently a phase difference between the two emerging beams. Finally, we must also choose the *angle of incidence*, or equivalently the *scattering angle* θ.

The appropriate combination of these three parameters will give rise to constructive or destructive interference. Once we have chosen some material with some specific chemical composition, the crystal lattice layers distance d is fixed. If we maintain the wavelength of our X-ray source as constant, the wavelength λ is fixed too.

This means that, if we observe the outgoing interfering waves at different angles θ, then we will see constructive interference (Fig. 35 left) or destructive interference (Fig. 35 right), due to the fact that they correspond to different paths of the light beams, giving rise to the respective type of interference.

This can be summarized by *Bragg's law*, which tells us the conditions for constructive interference from successive crystallographic planes:

$$n\lambda = 2d\sin(\theta)$$

Here n is an integer. For each integer (n=1, 2, 3,...) we are able to calculate each angle θ where the maxima will occur.

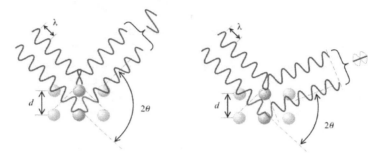

Fig. 35 Bragg diffraction: constructive and destructive interference due to reflection of X-rays on crystal lattice layers.

So, if we look at different scattering angles at the same time, for example on the screen of a photographic plate or a CCD sensor placed in such a manner that several beams with several angles are captured throughout a surface, then we will observe a very interesting diffraction pattern, as in Fig. 36.

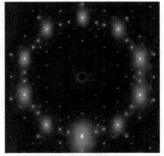

Fig. 36 X-ray diffraction and interference from a crystal lattice (scandium-zinc crystal—Ames Lab, Alan Goldman).

Here we have a beautiful symmetric pattern obtained by X-ray interference through a crystal. From these patterns, physicists and chemists are able to infer the crystalline atomic structure of substances. Therefore, Bragg diffraction allows us to reconstruct the geometrical displacement of atoms in the lattices, in some sense; we might say that it is like a microscope that indirectly furnishes images of the chemical structure of substances.

Admittedly, this is a simplified model. In practice, other planes also contribute to the overall effect, though to a lesser degree, and these will cause many more small peaks and complex patterns. But for our purposes here, this model suffices to illustrate the principle of Bragg's diffraction.

But what does this have to do with QM? Well, a lot.

First of all, it is necessary to dismiss a common misconception. Since at a macroscopic scale the double slit experiment displays light as a wave, while the photoelectric and the Compton effect vindicate the particle nature of light in a microscopic domain, one might be tempted to believe that, microscopically, light is in fact a particle—in the sense that it might appear and behave like a wave only at human scales, while close observation at the scale of molecules, atoms, and particles reveals it to be a particle and just like a water wave to be microscopically made up of tiny H_2O water molecules.

Not so. Bragg diffraction and interference phenomena act at the size of an atomic crystal lattice layer and cannot be explained with classical particle behavior. Considering light merely as a wave, Bragg father and son could explain the observed phenomenon, just as Young did using the double slit experiment. T. Young would have been delighted to see the wave nature of light confirmed also at microscopic scales.

But there is much more! One question that must arise at this point is whether this ambiguous twofold wave-particle manifestation is something inherent only to light. Since light seems to present itself as a wave or a flux of particles according to whichever kind of experimental setup one chooses, does this duality also hold true for material particles with a mass, like electrons or protons?

It might sound unreasonable and counterintuitive to think of material particles like an electron, which we imagine as a tiny chunk of matter, to behave like a wave as photons do. But an attempt to categorize them under the same umbrella came from the French physicist Louise de Broglie, who was able to synthesize in a unique theoretical concept the corpuscular and wavy nature of material particles. L. de Broglie realized that there is no logical and physical reason to believe that material particles with a mass could not have this double corpuscular and wavy nature as well as photons do. He imagined all the particles—including those which make up objects like atoms, molecules, and every material body in the Universe—no longer as particles or waves but as something which contains both, conceptually speaking: the so-called 'wave packet'.

Fig. 37 Louise de Broglie (1892-1987)

This was the famous *de Broglie hypothesis*, framed in 1924. The idea was to imagine every object being a traveling wave, but with the special property of being localized in space. If we see things in this perspective, we can consider any particle (with or without a mass) as having a specific wavelength, with the amplitude of the wave having a maximum in a region of space and which we intuitively think of as being classically the position of the particle, but with this amplitude decaying quickly from the center of the peak in all directions. In Fig. 38 two arbitrary examples of wave packets are shown: on the left a broader wave packet and on the right a smaller one. The two differ in their wavelengths.

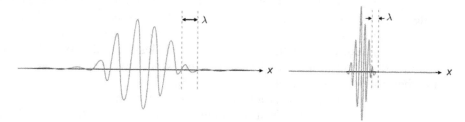

Fig. 38 Traveling 'wave packets' according to the de Broglie hypothesis.

If you look at these wave packets in *Fig. 38* from a large distance compared to their wavelengths (say, from a distance of ten meters from this book page), you may approximately consider them particles, since their wavy nature is no longer discernible to your eye. But if you look at the wave packets at a distance of the order of their wavelengths or shorter, they appear to be waves. In this sense we may speak of *'particle wavelength'*: an expression which otherwise, in the classical context, would have seemed nonsensical.

A gentle warning is compulsory here. One might be tempted to interpret the size of the wave packet as the size of the particle or the object under consideration. However, we shall see that this intuitive understanding is not correct. For the time being, just take the wave packet for what it is: a mathematical abstract construct that helpfully explains the observed phenomena of the experiments.

L. de Broglie quantified all this with a very simple and neat formula, which relates the wavelength λ of a particle, or more precisely the wavelength of its corresponding wave packet, as being inversely proportional to its momentum p, which in CM is defined as the product of its mass m times its velocity v:

$$p = m\,v. \qquad Eq.\ 5$$

De Broglie's wavelength is then defined as

$$\lambda = \frac{h}{p} = \frac{h}{m\,v}, \qquad Eq.\ 6$$

where h is, as usual, Planck's constant. The right-hand side of the equation holds true only for material particles with a nonzero mass and is correct only if we ignore relativistic effects. On the other hand, the term $\frac{h}{p}$ is more general and boasts universal validity, for photons as well as for material particles, and is exact also in relativity. It associates a wavelength with any kind of particle, independently if it has a mass (say an electron, proton, or a neutron), which moves with speeds less than that of light, or if it is massless, as in the case of photons, and which, in vacuum, always has no other possible speed than c.

This can be shown with a (not too rigorous) proof which highlights where this equation comes from. From Einstein's mass-energy relation (Eq. 4) we know that $E = mc^2$, which can be rewritten as $m = \frac{E}{c^2}$. If we take seriously the idea to put photons and material particles on the same footing, we can replace the mass in Eq. 6 with the latter expression and the speed with that of light ($v = c$) to obtain:

$$p = \frac{E}{c} \qquad Eq.\ 7$$

This expresses the momentum a photon of energy E carries. Photons, as all other particles, are able to carry and transmit certain amounts of momentum to other particles. Recall how photons, even when not visualized as material particles themselves, are able to give a 'kick', so to speak, to material particles, as we already saw in the case of Compton scattering. Now, recall also the ever-present Planck equation in the form we wrote in Eq. 3: $E = h\nu = h\frac{c}{\lambda}$. Rewriting all of this in terms of λ, we obtain $\lambda = \frac{h}{(E/c)} = \frac{h}{p}$, with the last replacement due to Eq. 7. This proves de Broglie's relation.

So, what de Broglie did was to put together the concept of the wave and particle into a single entity: the wave packet, the wavelength of which is determined by its momentum. In the case of material particles—not photons but particles with a mass moving slower than the speed of light—it is customary to speak also of '*matter waves*'.

The matter wave of the de Broglie hypothesis also sheds some light on Bohr's atomic model. In fact, if we understand electrons not as tiny localized point-like particles which orbit around the atomic nucleus, but as matter waves, that is, waves distributed concentrically around the orbit, it becomes quite natural and intuitive to understand why the energy levels of atoms must be quantized.

They must have discrete levels for a simple reason. If this matter wave has a wavelength which does not match and reconnect to itself again at the same point along the orbital path (see the diagram of Fig. 39) and if the circumference of the atomic orbit of the electron (the circle) is not exactly an integer multiple of the electron's wavelength (the wavy line), then constructive interference with itself is no longer possible, and the wave packet distributed around the atomic nucleus cannot build up as a standing wave structure (recall the standing wave we already discussed for the blackbody cavity).

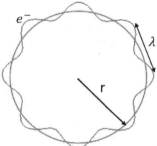

Fig. 39 Electrons as matter waves in Bohr's atomic model.

Here the principle is the same, but instead of reflections of a wave on a cavity wall, the wave circulates along an atomic orbit and for the same reason allows only for specific oscillating modes to exist. The classical notion of orbital velocity of the electrons around the nucleus is fixed by the fact that the electrostatic attractive forces between the positive charge of the nucleus and the negative one of the electron must be balanced by what we may call the centrifugal forces, loosely speaking, which keep the electron in orbit and which are given by its distance from the nucleus. But this distance cannot be an arbitrary one; the velocity of the electron, which is the wavelength of the electron (due to de Broglie's relation), must match just that orbital distance where constructive interference occurs and is equal to just that speed which is enough to keep the electron in a stable circular orbit.

It is possible to calculate this, and indeed it turns out that the electron will distribute itself naturally only on well-defined, discrete orbits. For the simple case of the hydrogen atom, this can be calculated with relatively simple algebra and CP concepts plus the de Broglie relation; in the end, only those matter standing waves are allowed to exist which match the quantized energy levels experimentally observed in the hydrogen atomic spectrum (see

Fig. 13 and *Fig. 22*). Bohr's atomic model added with the de Broglie hypothesis seemed therefore to gain credibility, since it explained naturally why atomic levels are quantized. Unfortunately, as we shall see later, it does not explain the spectra of all the other elements.

So, let us turn back to Bragg diffraction and clarify what we really mean by 'matter waves'. Since we can conceive both material particles, like electrons, and light waves as matter waves, and since we observed the Bragg diffraction for EM waves, as was shown for photons in the X-ray spectrum, the question at this point is: Does the Bragg diffraction hold equally also for material particles? Is it conceivable that tiny chunks of matter like electrons, neutrons, protons, or whichever massive particles can be diffracted by the lattice of a crystalline substance and display the nice interference patterns as has been observed with X-rays?

To check this idea one must send no longer photons, but instead material particles—say, electrons or neutrons—into the crystal lattice by modulating its energy, that is, its momentum (its speed). In fact, by doing so, one modulates, according to the de Broglie relation, its wavelength as well. If the wavelength of the wave packet matches the size of the crystallographic planes' separation, one should expect the interference patterns.

This experiment was conducted in 1927 by the American physicists L. Germer and C. Davisson, and independently also by the British physicist G.P. Thomson.

Fig. 40 From left to right: Clinton Davisson (1881-1958), Lester Germer (1896-1971), and George Paget Thomson (1892-1975).

Passing a beam of electrons through a thin film of metal, they observed how electrons are subjected to the exact same interference laws that photons are. What they found were indeed the Bragg diffraction patterns, like those of Fig. 41.

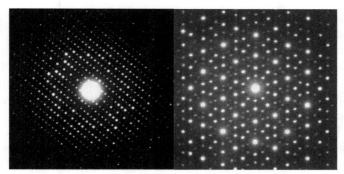

Fig. 41 Bragg diffraction and interference patterns from a crystal lattice of electrons (left) and neutrons (right).

So, when subjected to appropriate physical conditions, what was imagined to be rock-solid chunks of matter displayed a wave-like nature. This was an experimental verification of the de Broglie hypothesis, which at this point was no longer a hypothesis at all but an experimental fact: The observed interference pattern corresponded exactly to those expected if we adopt the de Broglie relation. This is a universal property of Nature—which, however, is not at all intuitive.

From the abstract mathematical point of view everything is clear and is nicely quantified and experimentally measured as described by theory. But from the more ontological perspective things are not that clear. The question is: What are these particles or matter waves really, at the microscopic level?

Are they particles, waves, something which should be conceived as a synthesis of both, or something else entirely? It is here that the philosophical journey of QM began, and this led then to the famous wave-particle duality, the uncertainty principle of Heisenberg, the concept of the wavefunction, and so forth, as we will explore in the following sections.

6. The wave-particle duality

Almost everyone has heard of the famous 'wave-particle' duality, but there are unfortunately many misconceptions surrounding it. One of the most frequent sounds more or less like this: 'A particle can be sometimes a wave or a particle, depending on the observer'. Apart from the fact that similar statements are already self-contradictory (if a particle *is* a wave, it is not a particle in the first place, eventually one should speaks instead of a particle that acts like a wave—that's why some physicists would like to eliminate any reference to 'dualities' from textbooks altogether), it is high time to debunk the 'observer myth' in QP. No quantum phenomenon needs a human mind or consciousness observing it to validate its existence, and there is no 'particle transmutation' in the sense that our naive intuition tends to suggest. It is no coincidence that we are presenting wave-particle duality to the reader after dwelling first on a historical and conceptual introduction in order to furnish the context with which to grasp the deeper meaning and implications of the theory. We do the same for Heisenberg's uncertainty principle in the next section.

We have seen that photons can behave either as waves or as particles, according to the context and the experimental arrangement. For instance, the photoelectric effect or the Compton effect gives one answer, and the double slit experiment and Bragg diffraction give the opposite one. In the latter case we have also seen how the wave character fits with material particles too, like electrons or neutrons.

However, we might ask at this point: When, how, and under which circumstances does the one or the other aspect arise? Is this a property that quantum objects acquire switching from one state to another? Is there somehow a moment in time where a somehow ill-defined entity is or behaves like a particle and afterwards morphs into a wave? If so, does this happen in a continuous fashion, or instantly? Or is there another possible interpretation and understanding which unites the two aspects of photons, as waves and as matter and energy?

In some sense we have already seen how, with the de Broglie hypothesis, it is possible to unify this disparity with the concept of the wave packet. Therefore, so far we have shed some light (pun intended) on this wave-particle duality, but the picture is still somewhat incomplete. Because, if we think about it carefully, nobody has observed a light wave in the way we can observe a water wave on the surface of an ocean. What we always observe are the effects of something that we imagine, by deduction after the fact, to be something that acts like a wave. The several experiments that hinted at the particle- or wave-character of light did *not* show that light *is* a particle or *is* a wave, but that it *behaves* like a particle or a wave. We should not confuse our human mental (sometimes even ideological) projections of how reality appears, with reality itself.

It is here where Young's famous double slit experiment comes into play again in its modern quantum version. It clarifies further some very important aspects and clarifies the nature of physical objects in QP.

So, let us turn back to the transverse plane wavefront of light (see Fig. 42), say, just a beam of light coming from a source which ideally is highly monochromatic, that is, having a single or very narrow range of wavelengths and colors (nowadays an easy task to accomplish with laser light).

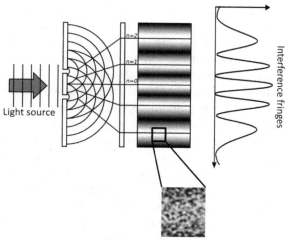

Fig. 42 The double slit experiment revisited.

We know that the slits should have a size and separation which should be comparable with that of the wavelength of the incoming light (something which, even for small wavelengths, can be achieved with high precision using modern lithographic technologies). When the plane wave goes through these slits it is diffracted and is detected with a screen at some distance, and the usual resulting interference pattern appears. These fringes can be easily photographed with a photographic plate or a CCD camera or whatever kind of sensor sensitive to the wavelength we are working with.

Through Young's experiment, the question about the nature of light seemed to receive a clear answer. We conceive of light as a wave since it behaves like a wave, due to displaying the interference phenomenon typical to waves.

The (more or less unconscious) logical background of this conclusion relies on what is called the 'duck test' of *'abductive'* or *'retro-ductive'* reasoning: *'If it looks like a duck, swims like a duck, and quacks like a duck, then it probably is a duck'*. This is what philosophers mean by 'counterfactual definiteness', that is, the ability to assume the existence of properties of objects even when they have not been measured directly.

However, upon closer inspection at the microscopic level, this turns out to be just an upside-down view of the larger story. How so? Because what we detect are always and only interactions between the light and the detector screen in tiny localized spots. What one observes locally is not some nice continuous shading of the intensity between the black and white fringes, but instead a random distribution of points, that is, of point-like localized interactions on the detector screen, just like pixels on a monitor. For instance, if we use an old conventional photographic plate as our detector screen, develop it, and then look at the photograph under a microscope (magnifying the region of the interference pattern in Fig. 42), we see many tiny white spots which correspond to the grains that, due to a chemical reaction, become white if a photon is absorbed.

After all, when we make a measurement, what we ultimately observe are particles, not waves. Be it a photographic plate, a modern CCD-sensor, or any other kind of detector which responds to an electric signal or just 'clicks' (like a photodiode or a photomultiplier), what we really measure are one or many dots, local point-like interactions which form an interference pattern, never a continuous pattern. Even if we were to build a detector with the highest resolution possible—say, every pixel is an atom—the same dotted interference pattern would emerge at smaller scales. This should not come as a surprise, since from what we have already learned about the blackbody radiation theory, we know that matter, and therefore atoms and molecules, absorb 'quanta' not in a continuous fashion but rather in the form of discrete amounts of energy.

Therefore, also in Young's experiment, we recover and find again the corpuscular nature of light: What seems to be diffracted at the slits and interferes on the screen behaves like a wave, but what is finally observed are photons hitting the screen. On one side, we have something we imagine to be a wave that goes through the double slits; but when we attempt to detect it, it inevitably shows up as a localized interaction.

In a certain sense we find ourselves again at the starting point. What is a photon *really*? If it is a wave, why, when, and how does it become a particle in order to be detected as a point-like structure?

Let us further inspect this state of affairs. If we would strictly maintain the particle picture, then we must assume that each particle must go through one or the other slit (see Fig. 43 left), and then travel further in direction of the detector screen and show up at one or the other fringe, that is, in two sets of piled-up particles detected only in two locations on the screen.

However, we know now that this is not the case, since we observe other lines appearing at the interference fringes as well. This naive picture is simply disproved by facts.

If we still believe that there are only particles, we are forced to assume that something like in Fig. 43 right must occur.

One has to assume the existence of some ill-defined 'wave' or 'force' or 'field' that 'guides' the particles along its path, so that they match the interference fringes. Indeed, this is an interpretation that is taken seriously by some physicists, with the so-called 'pilot wave theory' (described further in our second volume), originally formulated by the American physicist David Bohm (1917-1992). But I strongly suggest you carry on reading and learn what so far is held as true, before jumping into the plethora of interpretations of QM; many of these speculations can boast no hard supporting evidence.

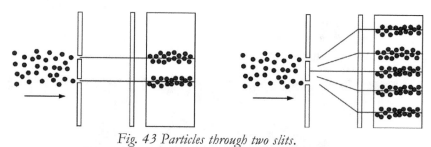

Fig. 43 Particles through two slits.
Left: What would be expected but is not observed.
Right: What is observed and cannot be reconciled with a classical 'particle-picture'.

In fact, let us take a further step in our reasoning and ask ourselves what happens if only one photon is sent towards the double slits. Do interference fringes arise also with a single photon? Or are they the result of the collective interaction of many photons? Or, which seems to be another possible alternative interpretation, does the single particle interfere with itself?

The latter question may seem at first meaningless, because we might object that a single photon could not be subjected to interference, since to observe these interference phenomena we need a collection of photons to build up as fringes on the screen. However, there is a way to test this hypothesis: We could try to deceive Nature by seeing whether and how the interference pattern builds up if we send only one photon after another and then wait to see how things develop over time. One after another means that we shoot a single photon at the two slits, wait until the single photon hits the screen and record its position, and only after that, we send the next photon, wait again until it is detected on the screen, and so on.

Experimentally this is not easy, but nevertheless possible to accomplish by making the light source extremely dim (for example, take a luminous source and dim it with some almost black foil), in such a manner that it emits only one photon at a time (and you should know by now that this is precisely what matter always does, if you recall what we said about the blackbody radiation and atomic transition spectra). Then the single photon is forced to go through one or the other slit. If it is a particle, as we imagine it being in our naive intuition, then it cannot go through both slits. If this particle travels through the first slit, then one expects to find it colliding with the detector screen in the region of the first fringe of Fig. 43 left. And when another particle goes through the second slit, then it is expected to hit the second fringe. So, since we are sending only one photon at a time, this seems to be a method to force Nature to manifest its corpuscular aspect.

This experiment proceeds as follows. We produce only very few photons at a time and then look at the detector screen after a short time interval. In Fig. 44a you can observe how only few photons are distributed in an apparently random manner all over the area where previously we observed the interference fringes.

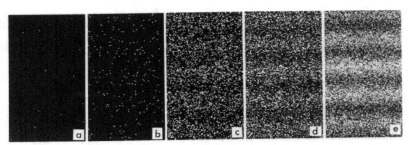

Fig. 44 Interference fringes appearing from a 'one-photon interference'.

So, there is no hint that interference occurs, and we are happy with that: We believe to have sent through the slits only one photon at a time and we observe only tiny white spots, as expected (even though the random displacement of the photons seems already to disprove the existence of only two lines). If we wait a little longer and collect more events on the detector screen, we see something like in Fig. 43b: Things look still clumpy, but now there definitely seems to be no trace of the two expected white lines. At this stage of the experiment, however, we still cannot discern clearly what the truth is. Then, after collecting a sufficient number of events on the screen, we begin to observe indeed the interference fringes, as in Fig. 44c. And, as we wait even longer and collect larger numbers of spots on the screen, the whole interference structure finally emerges (Fig. 44d and 44e).

We must conclude that, despite having dimmed our light source to such a degree that it emits only one photon at a time, the wave-particle duality does not disappear. The interference phenomenon is not a collective phenomenon whereby we might imagine many light particles somehow interacting among themselves to produce the interference fringes. Single photons obey the interference law! We might interpret the overall result as if the single particle goes through both slits at the same time, as a sort of 'ghostly wave' that is subjected to the laws of diffraction and interference, but then, when its position is measured, suddenly collapses to a single point at the detector screen. There it appears as a little localized spot. The kind of classical model of reality we imagined in Fig. 43 left is no longer tenable at all. It is simply wrong.

What this shows instead is that as long as light is not observed, it travels and expands throughout space as a wave; but when the act of measurement is performed, it chooses to place itself on a specific location. We see that something that seems to behave as a wave, suddenly 'collapses' to a point. We will never see the wave and particle nature at the same time.

An important aspect always to keep in mind is that nevertheless, the process is completely random in the following sense. Were we to look at this interference pattern only at a small local scale (again, with a microscope as in Fig. 42) and without knowing how the rest of the interference picture is made, there would be no way to know if the single white dot on the detector screen is the result of an interference phenomenon through the two slits or a single photon coming from a source with or without the slits in between. It is only when we look at the global picture that we can infer that interference must have taken place. In this sense we say that QP is governed by purely probabilistic laws. We can say that there is a certain probability that a photon is detected in a specific position, but we will never be able to tell in advance what this single event will turn out to be. We can only predict what will happen statistically, after a high (statistically significant) number of events have taken place. When such a number is collected, the overall disposition of the photons on the detection plane always obeys the classical interference laws, with its typical geometrical disposition of the dark and white fringe pattern. The single event, however—the displacement of a single photon on one or the other fringe—is a purely probabilistic event. There is a higher probability that the single photon will be detected on the brighter fringes, but there is no way, no method at all, not even in principle, to predict in advance which of the fringes the single photon will be detected on.

This hints at a quite different conception than what we intend in CP with the words 'randomness', 'chance', and 'coincidence'. In CP the side of the die that faces up after having been tossed is a random result, in the sense that the physical process which determines how the die rolls on a table is so complicated and practically impossible to calculate that the outcome is unpredictable. But in principle, if we would know everything about the die, the forces that act on the die, how it is thrown on the table, every detail about the table itself, and the precise initial conditions of the system, we could ideally predict the outcome with a supercomputer that knows everything about the laws of CP. In principle, that would not be impossible.

This classical determinism has its roots in a way of thinking which is nicely summarized by the so-called *'Laplace's demon'*. P. S. Laplace was a French mathematician and astronomer of the nineteenth century who reasoned as follows.

'We may regard the present state of the Universe as the effect of its past and the cause of its future. An intellect which at a certain moment would know all forces that set Nature in motion, and all positions of all items of which Nature is composed, if this intellect were also vast enough to submit these data to analysis, it would embrace in a single formula the movements of the greatest bodies of the Universe and those of the tiniest atom; for such an intellect nothing would be uncertain and the future just like the past would be present before its eyes.' [4]

Fig. 45 Pierre-Simon Laplace (1749 –1827)

This determinism is (more or less consciously) still the leading paradigm in science. If someone (i.e., the demon) knows the precise position and momentum of every particle in the Universe, its dynamical evolution for any given time could be calculated with certainty from the laws of physics. But we will see repeatedly during this journey through the weird world of QP that this is an assumption that can no longer be considered true in the quantum domain. Even if we know everything about the system and the initial conditions of a single particle, the single event will remain forever unpredictable. Not because of our ignorance, as Laplace believed, even if we have a 'demon' which knows everything and is able to calculate quickly and precisely enough. It seems that in QM things are not random in the classical sense, but this randomness is an inherent aspect of Nature. We shall take up this issue again in the following chapters.

If you have followed what we learned about the electron and neutron Bragg diffraction, it should not come as a surprise that the wave-particle duality is something that holds for material particles as well. In fact, the double slit experiment which Young performed with light produces exactly the same results if we use electrons or other material particles instead of photons. Therefore, the only thing we can say for sure is what can be represented in Fig. 46, which is a different attempt to model the observed phenomena.

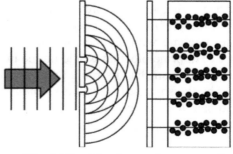

Fig. 46 Double slit experiment according to QP

Before the detection of anything, we should only imagine waves or matter waves in the case of material particles: first, a plane wave hitting the screen and, after the two slits, the two concentric wavefronts. Once these hit the screen they interact with it according to a probability amplitude dictated by the interference amplitudes and suddenly 'collapse' onto a single spot at the detection site. This is what the wave-particle duality is about. The wave-particle duality is about 'something' that propagates and spreads out throughout space behaving and interfering like a wave, but when asked about its whereabouts, answers as a point-like interaction. More than that, we can't say.

The question might arise if all this applies only to tiny elementary particles, as electrons, or if it holds also for molecules or even larger material objects? There is no limit in the principles of QM that prevent the de Broglie wavelength relation to be equally valid for any object with any mass m. In fact, it could be shown that this is indeed the case. Interference patterns arising from a beam of large organic molecules diffracting at single, double, and triple slit material masks could be observed experimentally. [5] In principle, also an elephant could be diffracted by a double slit device and interfere with itself. Why do we (fortunately) not observe this? Because the aperture of even the smallest slit we can manufacture is still huge compared to the extremely tiny de Broglie wavelength of macroscopic bodies. The elephant's wave packet will never suffer diffraction and interference effects. This should also make it clear how the de Broglie wavelength has not to be confused with the measure of the size of an object. If the size of a material object grows with its mass and becomes macroscopic, its de Broglie wavelength decreases down to sub-microscopic scales.

So, at this point we should be getting accustomed to the idea that there is neither a particle, molecule a 'chunk of matter' nor a wave at all, but only a physical phenomenon which we should not even attempt to visualize with our human intuitive macroscopic understanding of the world. The nature of matter can be described only in terms of probability laws.

Note also how we coincidentally are talking about the wave nature of photons or matter without detecting and measuring any wave at all but inferring its existence only indirectly, a posteriori—that is, after the measurement of its particle nature, by looking afterwards at the interference pattern, which consists of several white spots. We talk of matter waves without seeing waves at all, quite the contrary. Paradoxically, we deduce the wave nature of matter only when several measurements of particle-like objects have been performed. As we shall see later on, this is a prelude to a typical situation which occurs frequently in QP: The kind of situation where we tend to reconstruct in our mind an objective reality, a model of something which is already in the past and no longer in the present, by looking at the effects it produced. We tend to believe there was a wave in the past, despite the fact that we observe only particles in the present time. It is just this logical reverse engineering, and the question of the extent to which we are allowed to resort to this kind of reasoning in QP, that has given so many headaches to generations of philosophers. Honestly, we really don't know what these objects *really* are which we call 'photons', 'electrons', 'protons', and 'neutrons'. We even are no longer sure if it makes sense at all to ask for a description of reality that is supposed to have an objective existence and ontology independent from our mental models.

The fact is that physical reality must be a bit more complicated and counterintuitive than we thought. We have to deal with phenomena that refuse to be forced into an objective ontological worldview made of classical objects independent from the act of measurement—at least, not in the classical Newtonian physics sense. In the world 'out there', there is something that produces effects which in turn become the causes of future events, and that tells us this: *'I am a particle when you look at me but, as long as you don't look at me, I will continue to behave like a wave'*.

From the abstract mathematical point of view there is no problem with this. We just make all of our calculations with our nice wave packets and wave propagation and interference laws, and we obtain as an answer that which is confirmed by the experimental results. From the ontological point of view, things are not so straightforward. It is not clear what QP is trying to really tell us about the physical world and whether it makes sense to speak about objective realities at all. This aspect has puzzled the best minds of science and philosophy in history and continues to do so into the modern day.

7. Heisenberg's uncertainty principle

Another famous principle that almost everyone has heard of at least once in their lifetime is *Heisenberg's uncertainty* (or *indetermination) principle*. Usually it is explained with a few sentences or with wrong analogies and people (sometimes even professional physicists) have a misled understanding of what it is about. Therefore, to get a clear picture and avoid misconceptions, it is instructive to further deepen diffraction and interference phenomena for the one, two, and many slits cases.

So far, we concentrated our attention on Young's double slit case. In that occasion we saw how, when a couple of slits diffract a transverse plane wave, the typical interference pattern is produced, made of white maxima and dark minima fringes on a detector screen. And we saw that this is equally true if we send only one photon at a time through the slits. After a sufficient number of photons have excited the single photosensitive molecules of the photographic plate or the CCD camera sensor, the interference fringes reappear again.

The question at this point is: What happens if we use only one slit? Or vice versa: What happens when we use many slits?

By the end of the seventeenth century and well before the birth of QM, the single and N-slits interference phenomena were well known and extensively studied. As every good textbook on electromagnetism will tell you, a single slit, or a pinhole, will produce an interference pattern, though a less pronounced one than a larger slit produces. We mentioned how Young first studied the diffraction at a pinhole before experimenting with double slits. It is an easy experiment that you can do by yourself—just take a piece of paper, poke a little round hole with a pin or needle, and look towards a light source. You will observe the image of the light source surrounded by several concentric colored fringes (the colors appear only due to the fact that, fortunately, the world we live in is not monochromatic).

For example, Fig. 47 shows how a plane wave behaves when one aperture, a single slit of size a, is about five times the size of the light's wavelength, that is, $a = 5\lambda$. Due to the diffraction that occurs whenever a wave encounters an object or a slit, especially when its size is comparable to the wavelength, the plane wavefront is converted to a spherical or a distorted wavefront, which then travels towards the detector screen.

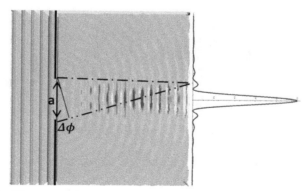

Fig. 47 Transverse plane wave incident on a single slit with aperture $a=5\lambda$.

In this case you see that, even though we are dealing with only one slit, some weak but still clearly visible secondary minima and maxima fringes can be observed.

An important point to keep in mind is to avoid a common misconception (which is frequently promulgated in some popular science books) which states that only two or more slits can produce interference fringes, whereas, for the single slit, interference effects disappear. This is not entirely correct. True, it is easier to produce more pronounced interference patterns with more than one slit (or pinhole); and for most applications, especially when the wavelength of the incident wave is much smaller than the size of the aperture, these effects can be neglected. However, strictly speaking, a single slit also produces small diffraction and interference phenomena.

An elegant explanation of how interference comes into being, also for a single slit, dates back to the French physicist A. J. Fresnel. He borrowed an idea

Fig. 48 Christiaan Huygens (1629 – 1695) and Augustin-Jean Fresnel (1788 – 1827).

from Huygens (hence the name 'Huygens-Fresnel principle'), according to which every single point on a wavefront should itself be considered a point-source of a spherical wave. Therefore, once the incoming plane wave reaches all the (virtually infinite) points along the aperture, every single point becomes a spherical wave point-source (see Fig. 49).

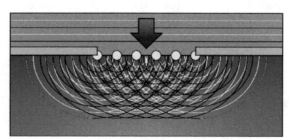

Fig. 49 The Huygens-Fresnel principle.

All these point-like sources along the slit's aperture emit at the same time their own spherical wavefronts which, however, when seen from a position on the screen, add up to produce an interference pattern. The reason for this is not so difficult to visualize. Since all fronts are initiated in different locations along the aperture they will also travel a different path length, which implies that they have different phase shifts when they overlap on the screen.

For instance, consider Fig. 47 once more, where we saw the two paths of the two sources from the edges. As in the case of the double slit, these two rays have a relative phase shift by an amount $\Delta\phi$ and, when superimposed together on the screen, they form a resultant intensity according to the interference laws. In this case, however, this holds not only for two waves, but for an infinite number of point-sources along the aperture. Fresnel, by making the appropriate calculations, was able to show that if one sums up all of the spherical wavefronts coming from the points of the aperture of the single slit and projects these onto all the points along the detector screen, then one obtains indeed the known diffraction and interference patterns.

If we repeat the same experiment with a slit that has a size close to the wavelength of our incoming wavefront, then we see that the interference fringes disappear (see Fig. 50).

Only when the size of the slit is equal to or smaller than the wavelength are the fringes definitely absent. This is because the slit is so small that only a single point-source can form a spherical wavefront with a wavelength equal to the slit size, and there can be no path difference and phase shift with some other source which could produce the interference pattern. However, diffraction has become very large instead, so that the photons will displace themselves on a relatively large area on the detector screen, according to a bell-shaped distribution called the '*diffraction envelope*'.

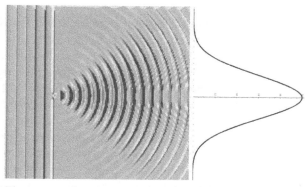

Fig. 50 Transverse plane wave incident on a single slit with aperture $a = \lambda$.

Let us take also a quick look at what happens if we change the number of slits maintaining a fixed size. *Fig. 51* compares three different cases. On the abscissa we have the angular position on the screen (angle of incidence θ between the center line of the diffraction plane and a point on the detector screen), while on the ordinate axis we have the usual intensity.

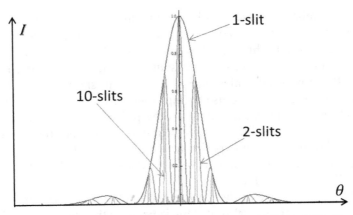

Fig. 51 Diffraction and interference pattern for 1, 2, and 10 slits.

The parameters that determine the angular dependence of the interference pattern are: first, the size of the aperture relative to the wavelength (here: $a=3\lambda$); second, the spacing d between the slits (here: $d=3a$); and, of course, the number of slits. The three curves represent respectively the 1, 2, and 10 slits diffraction cases. The intensities have been normalized in all the cases to unify.

For the one slit case you see there are some weak but clearly discernible secondary lateral peaks. They reduce almost to the diffraction envelope.

For the two slits, as in the case of Young's double slit experiment, we obtain more pronounced fringes. You can see how the one slit pattern 'envelops' the two slits pattern. The interference figure of the single slit pattern is split into several more fringes. However, notice that it would be incorrect to say, as you might hear frequently, that when we switch from the double slit to the single slit case, interference phenomena disappear. That is, in general, not the case. What happens is that we return to the envelope of the single slit which contains many fewer fringes, but still might have some other interference fringes too (and in this case it does). Again, interference is not a phenomenon specific to the double (or more) slits experiment. Interference does not disappear if one slit is covered; it merely becomes weaker than it is with more slits.

Finally, in the case of 10 slits, the two slits curve turns out to be the envelope of the 10 slits curve. So, you can observe how this is a more general trend and phenomenon which results from the interplay between diffraction and interference. In fact, generally, the N-slits fringes and their spacings arise due to this combined effect between diffraction and interference.

These were only a few examples to outline, at least intuitively, how wave interference works. Programmers who might be interested in recreating the infinite possible combinations of interference figures can find the general formula of the intensity function in Appendix A V.

We have seen in this section, especially with Bragg diffraction, how diffraction and interference phenomena determine the wave-particle duality not only of light particles, photons, but more generally of all particles— including material particles that have a mass and that behave like waves, even though we think of them as localized pieces of solid matter.

A question we might ask is: What happens with a particle if we want to know its precise whereabouts in space? For example, let us determine the precise position of a particle by letting it go through a tiny pinhole, as Isaac Newton did with photons in his investigations of the nature of light. If a particle goes through that single little hole on a piece of paper, we are authorized to say that we can determine its precise position in space. In fact, this can be done, but at a cost. Because, on the other hand, we know that due to the wave-particle duality, we can't forget the particle's wavy aspect. When a particle, also a material particle, goes through this pinhole, it will likewise be diffracted and afterward position itself on the detector screen according to an interference pattern.

If, instead of dealing with slits, we take a tiny round hole of a size comparable to that of a few multiples of the wavelength, we obtain circular interference fringes, like in Fig. 52. This is an intrinsic and unavoidable effect for all types of waves.

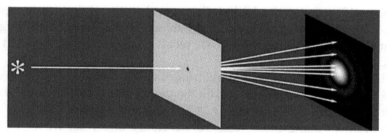

Fig. 52 The pinhole as a detector of a particle's position can't avoid interference.

If particles must be linked to a wave, according to the de Broglie relation, conceiving of it as a wave packet, we will always have the interference fringes, even with only one hole or slit and even with only one particle. (Recall the single photon experiment.)

Fig. 53 compares the two cases in which the pinhole determines the position of the particle with an aperture of $a=2\lambda$ (high precision) and $a=20\lambda$ (low precision), with λ, as usual, the wavelength of the photon or, in case of a matter wave, the de Broglie wavelength.

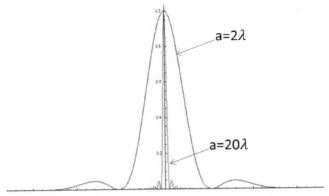

Fig. 53 Interference figure for the small and large pinhole.

If the pinhole is small, while it will determine the position of a particle with higher accuracy, it will also produce a relatively broad diffraction pattern that renders the position of the photon on the screen uncertain. We can know in a relatively precise manner where the photon or matter particle went through the piece of paper, inside the space constrained by the pinhole's aperture. But it will be displaced laterally on the screen anyway due to diffraction and interference phenomena.

Of course, with a single particle producing a single spot on the screen, no interference figure is visible. However, as we have learned with the single photon diffraction at the double slit, the probability of finding this spot is in one-to-one correspondence with the intensity of the interference fringes that many particles produce. Moreover, recall also that we can't predict where precisely this spot will appear.

If, instead, the pinhole is large, fringes will become less pronounced. We will know where the photon will hit the screen with relatively good accuracy (the spike for $a = 20\,\lambda$ in Fig. 53), which means it 'felt' only a tiny displacement along the screen. However, by doing so, we will lose our capacity to determine where exactly the particle went through the pinhole, as it is no longer a pinhole at all, but a large hole. There is no way, never ever, not even in principle, to obtain the precise measurement of the particle position and at the same time avoid the production of interference fringes (or interference circles, as in the case of a circular aperture or large diffraction effects). One will always obtain a more or less pronounced bell-shaped or peaked distribution of white spots on the screen. This is not because we don't have a sufficiently precise measurement apparatus but because it is a consequence of the intrinsic wave-nature of particles. It is a universal law of Nature, according to which it is hopeless to believe that we can pass a wave through a slit and not observe any interference and diffraction phenomena.

So, if you have followed thus far, it should be easy at this point to understand the essence of Heisenberg's uncertainty principle. Werner Heisenberg was a German physicist who, around the 1930s, asked himself if and how there was a way to precisely determine the position of a particle.

This was one of the most pressing questions for which to find an answer because, after all, in CP everything is expressed, analyzed and described in terms of space, time, forces and particles following accordingly precise trajectories. If we are no longer able to describe the position in space of a particle, we are substantially blowing up

Fig. 54 Werner von Heisenberg (1901-1976).

almost all known physics which, until then, relied completely on particles, interacting through forces with each other in space and time. Heisenberg reasoned as follows.

Let us suppose that we want to determine, with high precision, the position of a particle, as in Fig. 55. (Notice that in this figure, the x-axis is the vertical one.) Imagine it comes from the left with a specific amount of momentum p. Recall that the momentum of a particle, its quantity of motion, in classical non-relativistic mechanics is simply given by the product of its mass times its velocity. (Generally, it is a vector quantity, which means it also has a direction, something we will not bother with here.) Let us also recall that, as we have discussed in the lecture on the de Broglie relation, once a momentum is given, in QP we can also relate to every particle as a wavelength as: $\lambda=h/p$.

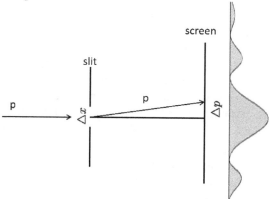

Fig. 55 Heisenberg's uncertainty principle explained using the single silt (or pinhole) diffraction experiment.

This particle, which we must intend as a transverse plane wave, is then incident on the diffraction plane with one slit (or pinhole) of size Δx.

Once it has passed through the slit, it will be diffracted and, shortly after, interact on the detection screen at a position according to a probabilistic law which the interference pattern determines, as you certainly know by now.

So, the particle, even without interaction from outside forces, even with no physical disturbance from the environment, and even by constructing an experimental setup which is perfect and ideal without any imprecision, will nevertheless be displaced on the screen (the vertical x-axis). This happens because of our attempt to localize the particle in that tiny region Δx. We can arrange our experiment as we wish, but if the size of the slit is comparable to the wavelength of the photon or to the de Broglie wavelength of the material particle, we will always and inevitably note that the particle, after it passed the slit of size Δx, will be displaced on the screen according to the laws of interference. There is no way to avoid this because it is an intrinsic interference phenomenon.

Now, this could also be interpreted as follows: The particle, once it has gone through the slit, will acquire an extra momentum, Δp, along the vertical axis. This does not happen because of the interaction of an outside force or, as we might imagine naively, by an interaction, deviation, or bouncing effect of the particle with the slit's edges because, in that case, we would observe a random distribution but not an interference pattern. This extra momentum Δp which displaces the particle along the detection screen is due only and exclusively to the wave nature of matter and light. We might interpret this also as a 'scattering' of the particle but we should keep in mind that this is misleading terminology, as there is no scattering force at all. No scattering interaction or forces from the outside are necessary to make this happen. Where does this extra amount of momentum Δp come from? It is simply the uncertainty we have about the particle's momentum in the first place. It is an inherent uncertainty of the properties of any particle due to their wave nature. This is the only possible conclusion if we want to avoid violating the principles of the conservation of momentum and energy.

The point is, we will never be able to determine with extreme precision – that is, with an infinitely small slit of size $\Delta x = 0$ – without blurring the momentum because, by doing so, we will inevitably diffract the plane wavefront, the wavelength of which is given by the de Broglie's relation. This will inevitably displace it according to a statistical law which reflects the diffraction and interference laws.

So, we must conclude that the smaller our uncertainty in determining the particle position (the size Δx of the slit), the larger the diffraction effects and, therefore, the larger the uncertainty over the momentum. (Δp becomes large in the vertical direction.) On the other hand, if I want to know the particle's momentum with small uncertainty (Δp small), we will have to open up the slit's aperture (Δx large) to reduce the diffraction. However, I will never be able to determine with precision both the momentum and the position of a particle at the same time. We have to choose whether we want to keep focused on one or the other; never ever are we allowed to obtain both. Again, this is not because we perturbing the system but because we are dealing with waves.

We have, however, interference fringes here, and in principle, the particle can displace itself very far from the center of the screen, especially if the slit becomes very small. However, this also happens with a decreasing probability. It is much more probable to find the particle hitting the screen at the central white peaks than at the little ones far from the center. For practical reasons, one must set a limit as a convention. Due to statistical reasons (which will be clarified later), one takes as convention the '*normal distribution*' which states that Δp is the width of the central fringe corresponding to 68.3% of all the particles it contains. That is because this is the central region with higher intensity and, therefore, that region where we have the highest probability of detecting a single particle. In fact, as we have seen with the single slit, the secondary interference fringes beyond the central one are usually quite small (smaller than in Fig. 55).

So, this means that, to a specific amount of Δx, we must expect to find also a specific amount of Δp – always and inevitably, by the laws of Nature. Heisenberg was able to show that the product of these two uncertain quantities is a constant and he summed this up with his very famous and extremely important inequality, which states:

$$\Delta x \cdot \Delta p \geq \hbar/2 \qquad \text{Eq. 8}$$

where $\hbar = \dfrac{h}{2\pi}$ (read "h-bar") is Planck's constant divided by 2π. This is a mathematical way of stating what we have said so far: If you force the value of the position of a particle to be tight, it will always have an uncertainty over the momentum of $\Delta p \geq \dfrac{\hbar}{2 \cdot \Delta x}$, that is, a precise measurement of the particle position will imply an uncertainty over its momentum. (A small Δx induces Δp to become large.) Vice versa, if you force the value of the momentum of a particle into a narrow range of values, it will always have an uncertainty over the position of $\Delta x \geq \dfrac{\hbar}{2 \cdot \Delta p}$.

This is an inequality which can be obtained in several ways. One closely related to this is via the Fourier transforms, that is, a mathematical tool developed by French mathematician Joseph Fourier (1768 –1830) who showed how every time series (a signal in time, such as a complicated traveling wave) can be decomposed into the sine and cosine functions of different frequencies. In modern QT, Heisenberg's inequality is obtained through a mathematical operator theory approach, which is mathematically more rigorous but somehow obscures the deeper meaning of Heisenberg's principle related to the wave nature of matter. At any rate, always keep in mind this little but very important inequality, as it will turn up more or less everywhere in QM.

Now let us analyze what interpretation Heisenberg gave to his own principle. He illustrated this with a famous 'thought experiment'. By 'thought experiment' (from the German 'Gedankenexperiment'), in science one means an ideal experiment which could be realized in principle and does not violate the laws of physics but can't be performed in practice due to technological limitations or other constraints. Heisenberg's thought experiment elucidates what nowadays is remembered as 'Heisenberg's microscope interpretation'. It is an interpretation that, unfortunately, several professional physicists have also adopted (probably to make it more understandable to the public or to first-year students). However, if you have carefully followed what we have said so far, you will be able to recognize that Heisenberg's interpretation is somewhat misleading. It went as follows.

Imagine (see Fig. 56) that we want to detect the position and momentum of a particle in a region Δx by using an incoming photon which hits the particle. This photon is then scattered toward some direction. We have already described something similar with the Compton scattering effect. Then, as was Heisenberg's reasoning, we can look to where the photon has been scattered to deduce the electron's position.

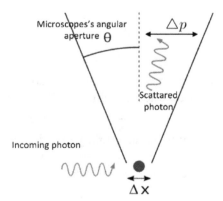

Fig. 56 The old Heisenberg's microscope interpretation.

However, according to classical optics, the larger the angular aperture θ of the microscope, the better it will resolve the position of the scattered photon and, therefore, deduce the position of the electron. Yet this angular aperture depends not only on the size of the microscope lens but also on the wavelength of the photon. To have high precision in determining from which direction the photon comes, that is, in determining the position of the electron, we must use a photon with a short wavelength. However, if the photon has a short wavelength (high frequency), it will carry more momentum because it is more energetic due to Planck's equation. Therefore, the kick that it imparts to the electron will also be larger. That is, if I want to observe the precise position of the electron, I must use photons which will produce a large recoil on the electron itself. This makes its position again indeterminate. On the other hand, if I want to more precisely determine the momentum of a particle, I'm forced to use photons with large wavelengths because only they have a small momentum and will not displace the electron too much through Compton scattering. Yet if I use long wavelength photons, the microscope optics will resolve their position with less accuracy. So, again, we must search for a tradeoff between the wavelength we use (the precision over the momentum Δp) and the microscope aperture (the ability to look at a precise position Δx).

In the macroscopic world, this effect does not play any role for practical purposes because objects are huge and not disturbed by tiny photons (even though, strictly speaking, it always holds). However, for atoms and even more for particles like electrons, a single photon can disturb them and scatter them away, preventing us from determining their properties. That's why Heisenberg's uncertainty principle is invoked to explain that in the microscopic world, a measurement always perturbs a system in such a way that the scale of the measurement's perturbation is about the same magnitude of the effect it produces and, therefore, renders the result imprecise if not entirely useless.

Heisenberg was able to show that, by reasoning in such a manner, making the appropriate calculations which resort to classical optics, he could indeed reobtain his famous inequality over momentum and position. Heisenberg's microscope thought experiment boils down to the following argument: We cannot determine the position and momentum of a particle at the same time because when we make an observation, we must inevitably interact with the object we want to observe (such as sending other particles of comparable mass or energy to the particle being observed). This inevitably disturbs the system, which then loses its original position and momentum that we sought to determine.

This interpretation, however, is in stark contrast to the wave-particle nature of matter itself. It is an interpretation that does not stand up to the tests of modern QT. The popular belief that Heisenberg's uncertainty principle is about the interaction between the observer and the observed object is wrong. Heisenberg's uncertainty principle is a fundamental law of Nature whose roots are in the wavy nature of matter and has nothing to do with the interaction between the measuring apparatus and the measured objects. Of course, the interactions and imperfections of measurement devices must always be taken into account in a real laboratory but these add further uncertainty to the intrinsic quantum fuzziness that is present a priori. In fact, we shall see later that, as strange as it might sound, it is nevertheless perfectly possible to set up experiments which make measurements without necessarily interacting with the system, and yet the Heisenberg uncertainty principle remains inescapable. It also holds in the case in which we are able to reduce to zero any interaction with the observed object. It is an inherent law of Nature that is independent of observational interactions.

At the time of Heisenberg, it was still legitimate to think of particles in this way. Heisenberg can be excused because several experiments which showed how his own interpretation must be revised came much later – some not even until before the 1990s, with the development of sophisticated laser and quantum optics devices. So, while the historical context and the lack of more experimental evidence justifies Heisenberg, it does not do the same with modern physicists. Nowadays, we can no longer stick with the microscope interpretation as a correct understanding of the workings of the uncertainty principle. This is no less fundamental than the geocentric model or a flat earth theory.

We should accept that Nature is telling us that we should never forget the wave-particle duality. When a particle goes through a pinhole, we must think of it as a physical process described by a transverse planar matter wave, which is diffracted like any other wave and produces a spherical wavefront – neither more nor less than in Fig. 57. Forget about classical understandings of particles which possess definite properties such as a position, and which move along precise trajectories and possess a definite momentum.

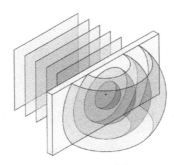

Fig. 57 The quantum interpretation of the uncertainty principle.

There are no positions or momenta which describe a particle. Instead, we must seriously consider these to be emergent qualities, not intrinsic properties. Particles do not at all have a clear and precise position and momentum as we imagine them to have at our macroscopic level and as our naive intuitive understanding wants to make us believe. QP seems to suggest that the physical objects we imagine as point-like hard material billiard balls are, instead, entities which are intrinsically somewhat fuzzy, with no sharp and well-defined boundaries and that travel throughout space as waves that eventually collapse.

This nature of particles is independent of the precision of our measurement apparatus and independent of the fact that we interact (or do not interact) by observing it. The position and velocity of particles are independent of what we know in the classical world in which we live.

Indeterminacy is not a matter of ignorance. Heisenberg's uncertainty principle is not a principle about a lack of information. It is not about particles whose whereabouts and speed we cannot determine. It is about physical entities that simply don't have anything to do with our anthropocentric imagination. This is what distinguishes classical from quantum physics.

8. The wavefunction and its 'collapse'

Let us return briefly to the de Broglie hypothesis. Here we had to deal with an abstract object, with a sort of conceptual synthesis that resembles quite well both a wave and the appearance of a particle. Every particle in QT, like the electrons, protons, photons, etc., can be represented by a wave packet, that is, a small localized wave that travels throughout space and that has a frequency (or wavelength, if you prefer), and a maximum central value that decays quickly on both sides (recall Fig. 38). It is a way to visualize a wave that nevertheless can interact locally as a particle and fits well into the wave-particle duality, where the quantum objects seem to behave as waves and yet as particles too.

Then, we have also seen that in QM we must deal with interference phenomena, which bear some similarities to that of the waves of CM. We introduced the concept of the phase between two waves, or a phase relative to an origin, and we saw how this is reflected in the formalism of QT.

Now, the mathematical function which describes a wave packet or any quantum mechanical description of the state of a system is called the *'wavefunction'* and is usually written with the Greek letter ψ (read 'Psai'). The wavefunction contains all the information about a particle or a quantum system in space coordinates x and in time t as ψ(x, t). This is, in fact, the conventional and standard mathematical description of waves in QM.

In section I.3 (with the help of Appendix A Id), we also found that the intensity of a wave – that is, the physical quantity that we really measure, say, the number of photons we measure arriving in a specific area on a detection screen – is given by the squared modulus of the wavefunction as,

$$|\psi(x)|^2 = \psi(x) \cdot \psi(x)^*, \quad Eq.\ 9$$

where, here, we consider only the change in the space, variable x, leaving out the change in time, variable t. Therefore, it is convention to depict a wave packet as in Fig. 58. You see the real part of a wavefunction (the wave with positive and negative values) and its squared modulus (the wave with only positive oscillations), which represents the effectively measured intensity. (The imaginary part is not shown here, for the sake of simplicity.) The Δx is taken as measuring the width of the wavepacket and corresponds to the uncertainty over the position of a particle due to Heisenberg's inequality.

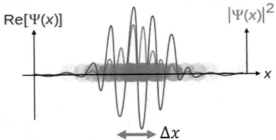

Fig. 58 The classical particle position (discs), the wavefunction's (positive and negative) real part, and its related probability density (always positive).

If you think of the density of dots, for instance, those forming the interference fringes on the screen of the double slit experiment, the number of photons hitting per unit area the photographic plate (or a ccd camera, or whatever kind of detector), the intensity is just proportional to the density of dots.

The registered number of particles on the screen per unit area (or, as it is represented graphically in this picture, like the transparency of the particle around an origin) can be taken to be proportional to the probability of finding the particle in that specific area.

Recall that while the overall disposition of many photons on the screen resembles that of the interference fringes, the site where the single particle will hit the screen is nevertheless a completely probabilistic process. The appearance of one photon as a dot on the screen is purely a statistical phenomenon. Therefore, one can regard the intensity – that is, the squared modulus of the wavefunction – as something proportional to a probability per unit area, that is, a *'probability density'*.

The *squared modulus of the wavefunction* is the master tool with which physicists describe probabilities in QM, that which connects theory with experiments and observations. This latter quantity is what allows an abstract complex mathematical function to correlate with reality, and which is described by the measurement processes that physicists perform in the laboratory.

We can introduce the following interpretation, which is due to the German physicist Max Born, and which is also called the *'Born rule'*: The squared modulus of the wavefunction according to Eq. 9 is a

Fig. 59 Max Born (1882-1970).

probability density to observe a quantum system in a specific state.

To clarify what this rule means, let us consider the graph of Fig. 60. Let us suppose that the curve represents the probability density to find a particle around an origin.

If the horizontal axis represents the position x where a particle can be found, the vertical axis is the probability density p(x) to find it at x. You will frequently find such kinds of bell-shaped curves in several textbooks; it is the *'normal distribution'* or *'Gaussian distribution'*.

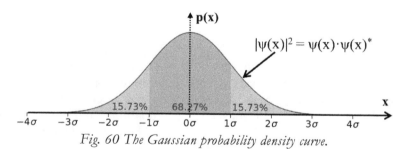

Fig. 60 The Gaussian probability density curve.

However, a probability density is still not a quantity that can be measured because it represents only an infinitesimal quantity. It makes no sense to ask what the probability is of finding a particle in a specific position, as the interval on the space axis, in such a case, would be zero. One can only say with what probability a particle can be localized in a specific spatial interval around the origin (say, the Δx we worked with in Heisenberg's uncertainty principle). The name 'density' is no coincidence. Just think of the density of matter. The material density function of a body describes how the density in it varies inside a volume. Yet this still isn't enough to tell us something about the amount of matter if we don't specify how large the volume is. Similarly, the wavefunction and its associated squared modulus, the probability density, describes all the possible states of a system but still does not tell us the probability of finding it in a range of possible states. What we must consider is the sum of the probability density function over a specific range, which tells us something about the effective probability P to find it in that range.

For example, for normal distributions, the spatial x-interval of $\pm\sigma$ ('sigma') is per definition the range that determines a 68.27% probability of finding a particle. σ is called the 'standard deviation' and tells us something about the dispersion of a set of data values. The probability of finding the particle far from the center drops off quickly. At $\pm3\sigma$, it is almost zero.

It might also be of interest to know that a mathematically rigorous approach to Heisenberg's uncertainty principle holds precisely when the uncertainty range over the particle's position and momentum is the standard deviation and, therefore, can be reformulated as: $\sigma_x \cdot \sigma_p \geq \frac{\hbar}{2}$.

In general, to obtain a probability P of finding the particle inside some spatial range $\pm x$, we must take the integral of the probability density over that interval –that is, the integral over the squared modulus of $\psi(x)$. An integral is the area under the graph's function which can be obtained as the sum over all the function's values at each point along the x-axis, in our case:

$$P = \int_{-x}^{+x} |\psi(x)|^2 \, dx.$$

Even if you are not accustomed to integral calculus, visualizing this is not so difficult. For instance, just think about how the mass of a body is given by the integral of a mass density function over all its volume. (If you struggle with the notion of the integral, check the appendix!)

And, of course, if you consider the probability of finding a particle anywhere, throughout the Universe, the spatial range must be infinite, taking x= $\pm\infty$ as the domain of integration. Then, obviously, you will get 100% probability – that is, the certainty of finding it somewhere, which mathematically translates into the integral giving unity:

$$\langle\psi|\psi\rangle = \int_{-\infty}^{+\infty} |\psi(x)|^2 \, dx = 1, \qquad \text{Eq. 10}$$

The expression in the brackets on the left-hand side of the equation is simply a more compact form to express the integral. It is quite common shorthand in QT that we will take up in more detail later.

More generally, the wavefunction applies not only to a particle's position but to any possible state. It is a calculation tool with which to also find the probability of finding a particle having a momentum or angular momentum (the rotational speed), an atom or a nucleus being in some energetic state, etc.

The takeaway message of all this is that QM is a purely probabilistic theory. This is due to its wavy nature. There is no way of knowing how a single particle will behave, not even in principle (except when the particle is in what is called an 'eigenstate', as we will see later). That is why, in QT, statistical tools play such an important role. We can compute the probability only to measure a value in some interval. In QM, it makes no sense to speak of a system that *is* in some state. It makes sense only to say that there is a certain probability that a measurement will furnish an outcome that we associate with some state at the instant of measurement. However, we are not allowed to say anything about its state before and after that measurement. The wavefunction is a state function only in the sense that it is a mathematical tool that tells us about the statistics and probability of finding a specific system in a specific state or range of possible states.

At this point, the question is whether the wavefunction is simply an abstract mathematical description of the statistical behavior of a system that simply updates our knowledge, or whether it represents a real object in the physical world. And, if it must be considered real, how can it be something that behaves like a wave and then turns out to describe particles? This is the so-called '*wavefunction collapse*', also called the '*state reduction*' or '*state projection*'. It is one of the typical quantum philosophical problems and will lead to the more general '*measurement problem*' we will discuss later.

It can be illustrated with what we already saw in the double slit experiment and what was depicted in Fig. 46 and summarized in Fig. 61.

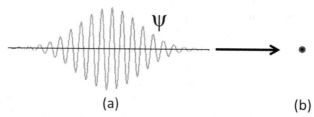

(a)   (b)

Fig. 61 The wavefunction 'collapse'-
a) Before the measurement; b) At the instant of the act of measurement.

First, a plane wave is diffracted by the slits. Then the two expanding concentric circular wavefronts emerging from each of the apertures travel towards the screen. These wavefronts can be described by the wavefunction

ψ. Once they interact with the screen, the interference fringes form but as a pattern of dots, each representing a point-like interaction at some specific position. Nothing is left of any wave.

This suggests an interpretation that tries (a bit desperately) to reconcile the two apparently opposite points of view. First, before the measurement, we must conceive of the single photon (or electron, or whatever particle we may be talking about) as a wave, or a wave packet which (still before the measurement, that is, before it hits the screen) continues to behave according to the diffraction and interference principles.

Nothing should be conceived of as a particle. When the wavefront goes through the slits, it is still a wave and no particle is supposed to go through one or another or both slits. (Or, only in this sense, we might be authorized to say that in QM, a single particle goes through both slits and interferes with itself.) Whereas, at the instant of measurement, in our case, at the instant when the wave interacts with the screen, it suddenly reduces, or 'collapses', to a point on the screen, manifesting itself again as a particle. However, before that instant when the act of measurement occurs, we must still consider it to be a wave with an extension in space, not as a single point. It was first the Austrian-Ungarian mathematician John Von Neumann who introduced the wave function collapse as a postulate and the description of a physical measurement in QM as a discontinuous, non-causal, irreversible and instantaneous process.

In fact, from the mathematical and formal abstract point of view, everything is clear: You simply calculate with the wavefunction and then take the squared modulus and obtain the probability density where you will find your particle. Then you integrate over a region of interest and obtain a statement about the probability of observing the system in some specific state (here, the particle to be detected inside as a spatial interval along the screen). The wavefunction collapse, or state reduction, is simply conceived of as an update of the information we have about the state of a system. Previously, we had uncertainty over a possible range of values that can turn out before a measurement and represented this lack of information with a statistical function. However, once the measurement is made, we know the exact value; therefore, there is no need to use the wavefunction. One gets only one measurement outcome and updates the knowledge one has about the state of the system. It is just like tossing a dice without looking at it. Before controlling the outcome, one has six possibilities to be realized. However, after one looks at the dice, only one of the six possible outcomes appears to be the truth. Therefore, the problem is not mathematical; we may say it is not even scientific. However, it is an interpretational problem.

Yet if that is a common interpretation in CP, it is not so in QP because, as the double slit experiment tends to suggest, the wavefunction seems to be more than a '*probability wave*' but, rather, something real propagating from the slits to the detection screen. (Otherwise, how can interference come into being in the first place?) Therefore, a question arises in the domain of philosophical speculation. From the ontological point of view, there is no consensus as to how we should interpret this 'collapse' and what kind of objective reality QM is describing after all. What kind of thing behaves as a wave unless one looks at it, and then collapses to a point instantly, in an infinitesimal interval of time? What is really 'out there'? Before the measurement, it seems that we have only a sort of ghostly entity whose wavy nature we will never observe directly. However, we can infer from the interference phenomena, and at the measurement act it nevertheless manifests as a completely different entity, a point-particle. And there is no way to intercept a morphing, a transformation from one entity to another.

After all, we are talking about a probability wave travelling through space. Yet the notion of probability is an abstract mathematical concept; it is just a mathematical function that mathematicians write with pencil on a piece of paper – a concept that we do not think of as an object having a concrete existence moving somewhere. One can see water waves on a lake, hear sound waves, perceive light, and see birds flying, but nobody has ever seen a mathematical function or the square root of -1 moving throughout space.

And yet, this is what QP seems to be about. It seems that not only must we resort to very abstract notions to describe the world but the world itself seems to be a sort of mathematical abstraction, reminiscent of a Platonic realm. Is anything physical 'out there', represented by ψ? Is the collapse of ψ only a statistical occurrence, or also a real physical one?

So far, nobody has been able to furnish a convincing and generally accepted answer to these questions. Most physicists simply do not bother and will tell you that they are too busy making calculations. They are happy that the math works. And, indeed, the mathematical foundations of QM were extremely powerful in furnishing predictions that could be measured and confirmed, and that led to the tremendously successful standard model of particle physics, atomic physics, solid state physics, etc. Yet from the interpretational point of view, not much progress has been made. This suggests that we still have not gone deep enough in our understanding of the world. For many years, physicists delighted themselves with calculations but with recent attempts to build quantum computers which require more or less direct answers to these questions, and after decades of unsuccessful attempts to discover a general theory of quantum gravity, that kind of theory that reconciles the gravity force with the electromagnetic and nuclear forces, these foundational issues have gotten more attention again.

9. The state vector, Schrödinger's equation, and atomic physics

We have seen that QM is a purely probabilistic theory. According to QT, events occur only with a certain probability but the single event can never be predicted with certainty unless the system is already prepared into a state called an *'eigenstate'* – that is, that kind of quantum state in which repeated measurements of the quantum system will always lead to the same result. Yet, in general, quantum events are purely random and we must resort to statistical reasoning and predictions as well as a statistical formalism and math. This is essentially what distinguishes CP from QM, the purely probabilistic and random nature of the latter.

As we already mentioned with Laplace's demon, in CP, if you fix the boundary conditions, a system will always evolve in the same manner, and it is determinstic and predictable, at least in principle. In QT, even if you shoot the particles from the same place, with the same speed toward the very same slits, you will always see them hitting different places on the detection screen. However, this does not mean that nothing can be said about the average behavior of a system, say, a system of particles, or its dynamic and temporal evolution. In this case, it is perfectly possible to describe it according to some statistical rules – for example, the statistics that describe the distribution of a large number of particles on a screen according to an interference fringes pattern. This also means that, in QM, the math describing the dynamics of a physical system must be very different. Instead of ascribing to a particle a single final possible state, one must resort to the wavefunction, which describes several possible measurement outcomes probabilistically, though the particle's initial condition is known.

Let us, therefore, go through a brief introduction to some formal aspects of QM. We will keep it a non-rigorous introduction in contrast to a complete course on QM but it is nevertheless necessary to digest some simple yet very important formalism and symbolism, as it will continue to emerge frequently here and there, and it is that kind of representation that you will encounter throughout all textbooks on QM.

So, let us turn our attention back to the concept of the wavefunction. We considered it to be continuous for the position in space of a particle, and its value, the density probability or the integrated probability function of the particle in space, does not change abruptly from one point to another. That is, the wavefunction $\psi(x)$ is a continuous function in x – or, technically speaking, the system has an infinite number of degrees of freedom.

We saw, however, that there can be other behaviors of quantum systems – for example, the energy state of atoms is quantized. Here, the wavefunction can no longer be a continuous function but must be represented by a set of discrete possible states of the system – for instance, as we have already seen in the case of the hydrogen spectrum (see Fig. 13) or in the F-H experiment. In these cases, we must write the wavefunction Ψ not as a continuous function in space but as a set of single states $e_1, e_2,.., e_N$, which can represent the discrete energy states $E_1,..,E_N$ or other dynamical variables (like the spin or angular momentum, as we shall see later). In the language of QM, these measurable quantities are called '*observables*', whereas each of these discrete states of a quantum system are called '*eigenstates*' (from the German 'eigen', which means 'self').

In QM, these eigenstates are defined on a Hilbert space. Without going too much into mathematical details, it suffices to say that a Hilbert space is an abstract vector space that is an extension of the real Euclidian space to a multidimensional space of complex numbers and complex vectors. If the Hilbert space has a finite dimension, it is the state space of a quantum system with a finite number of eigenstates. Otherwise, it is infinite – as infinite as the number of its degrees of freedom. Another important difference in a 'space' as we conceive of it in school geometry is that the elements of a Hilbert space are no longer points and coordinates but, rather, functions, with the wavefunction being the primary object on which to act.

The eigenstates $e_1, e_2,.., e_N$ form the '*eigenbasis*' of the Hilbert space, and the overall possible states of a quantum system are represented by a vector defined on this eigenbasis, called the '*state eigenvector*' or just '*eigenvector*' or simply the '*state vector*' as (in Fig. 62 a simple case for a system with three possible eigenstates is shown):

$$|\Psi\rangle = c_1|e_1\rangle + c_2|e_2\rangle + .. + c_n|e_n\rangle, \quad \textit{Eq. 11}$$

where $c_1, c_2,.., c_N$, are complex coefficients which are calculated and interpreted as follows.

Because we are no longer dealing with a continuous function but, instead, with a state vector in which each element corresponds to discrete eigenstates (think again of the 'eigenenergy' spectrum of the hydrogen spectrum), the probability density function of Eq. 9 must also be replaced by discrete probabilities. In general, the probability of finding the system in one or another eigenstate is not the same but the set of numbers which represents it must obviously sum up to unity (that is, certainty that the system will be found in at least one among all the possible states). The information regarding with what probability each of the eigenstates will be observed after a measurement is carried by the complex coefficients $c_1, c_2,.., c_N$.

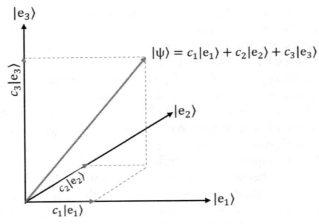

$$|\psi\rangle = c_1|e_1\rangle + c_2|e_2\rangle + c_3|e_3\rangle$$

Fig. 62 Quantum state vector in a 3D Hilbert space.

The Born rule still holds: The probability P associated with the i-th eigenstate e_i is the squared modulus of the i-th complex coefficient c_i :

$$P(e_i) = c_i \cdot c_i^* = |c_i|^2 \quad (i=1, 2.., N)$$

The notation $|\psi\rangle$ is a common notation for vectors in QT. State vectors can be written according to the 'Dirac-notation'. P. Dirac was a British physicist who made several contributions to QT. We already saw how he wrote the kind of integrals we encountered with Eq. 10 as brackets, like $\langle a, b \rangle$. Each bra-ket is a product (more precisely, the inner product) of a 'bra-vector' $\langle a|$ (a row vector whose elements are the complex conjugate of a) and a 'ket-vector' $|b\rangle$ (a column vector). As you can see, Dirac wasn't lacking a bit of weird lexical fantasy, though this turned out to be extremely useful. For example, the general 'bra-a ket-b' is:

Fig. 63 Paul Adrien Dirac (1902-1984).

$$\langle a|b \rangle = (a_1^* \cdot a_2^* \cdots a_N^*) \cdot \begin{pmatrix} b_1 \\ \cdots \\ b_N \end{pmatrix} = a_1^* b_1 + a_2^* b_2 + \cdots a_N^* b_N. \quad Eq.\ 12$$

This is called an 'inner product' and is just an extension of the familiar 'scalar product' or 'dot product' and that some might recall from their high school algebra, but extended to Hilbert spaces where vectors have complex entries.

With Dirac's notation, many calculations are made formally simpler. For example, if

$$\langle e_i | = (0 \cdots c_i^* \cdots 0)$$

and

$$|\Psi\rangle = \begin{pmatrix} c_1 \\ \cdots \\ c_N \end{pmatrix}$$

then the probability of finding the i-th eigenstate can also be written in the bra-ket notation as follows:

$$P(e_i) = |\langle e_i | \psi \rangle|^2 = |c_i|^2 \; ; \; \sum_{i=1}^{N} |c_i|^2 = 1 , \quad (i = 1, 2, \ldots, N) \quad Eq.\ 13$$

Again, with the right-hand-side sum of all the modulus squared complex coefficients, each representing the probability of the i-th eigenstate to be measured, must obviously sum up to certainty (or, to put it another way, the sum of the probabilities of all possible events gives unity).

It is just a symbolic formalism introduced by Dirac (and later developed further by others, such as the mathematician John von Neumann) with which one represents the probability of observing a system with a number of possible discrete states.

It is also customary to state that *Eq. 13* is the modulus square of the *'projection of the state vector'* ψ onto eigenstate e_i. (One can interpret the dashed lines in Fig. 62 as 'projections'.) It is the measurement act that projects the state vector onto one of its possible eigenbasis vectors. This is analogous to the wavefunction collapse we saw in the case of a continuous wavefunction. There, the act of measurement collapses the wavefunction describing the particle position in a continuous space on one of the infinitely possible states. (After all, the position of a particle is also a 'state'.) Here, it projects the state vector onto a discrete spectrum of possible eigenstates. This description of quantum measurements derives directly from experimental observation and was first formalized by von Neumann, also known as *'von Neumann's projection postulate'*.

Moreover, once the quantum system has been placed into one of these eigenstates, in QM jargon one says it has been *'prepared'* into a specific state 'i', and if it does not interact with the environment suffering due to external interactions' state reduction, every subsequent measurement will always deliver the same result: the system being in state $|e_i\rangle$.

Note how the sum of all the probabilities adding up to unity is represented in Dirac's notation neatly as:

$$|\langle\Psi|\Psi\rangle|^2 = (c_1{}^* \cdots c_N{}^*) \cdot \begin{pmatrix} c_1 \\ \cdots \\ c_N \end{pmatrix} = \sum_{i=1}^{N}|c_i|^2 = 1 . \qquad Eq.\ 14$$

If we want to deal with real-world measurements, we must specify which information from this mathematical construct (the wavefunction or state vector) we want to extract – that is, which physical quantity, and more precisely which '*dynamical variable*', we want to measure (including, for example, position, momentum, angular momentum, energy, etc.). Mathematicians and physicists define an '*operator*', which, as the word says, 'operates' on the wavefunction or state vector to obtain a specific piece of information associated with the observable.

Examples of mathematical operators that every one of us knows are the addition (+), subtraction (-), multiplication (.), and division (:) operators. The multiplication operator acts on two quantities, a and b, to obtain another quantity c as a·b=c. Similarly, one can conceive of an operator O acting on a state vector $|\Psi\rangle$. If we consider this operation to be a formal equivalence to an act of measurement of a dynamical variable, the observable of a system, then the output of this operation should tell us something about the outcome of this measurement, a scalar quantity λ, a number, which represents the expected value from this measurement, the so-called '*eigenvalue*':

$$O|\Psi\rangle = \lambda|\Psi\rangle .$$

λ is the outcome of a measurement or, more broadly speaking, simply an 'observation'. This equation is therefore called the 'eigenequation'. It is the mathematical procedure we must perform to make predictions once we know the experimental setup and boundary conditions.

Stated a bit more precisely, one should say that each dynamical variable (e.g., position, momentum, angular momentum, energy, etc.) is associated with an (self-adjoint) operator O, the observable, which acts on the state vector of the quantum system $|\Psi\rangle$ and which furnishes a scalar quantity, the measurement value, called the eigenvalue. Self-adjoint means that the complex conjugate of the operator is again the operator itself; however, these are mathematical aspects that are not very important for our purposes. Simply remember the connection between the classical notion of dynamical variables and its counterpart in QM as operators, the observables.

To avoid a possible source of confusion, let us state this again from another perspective. Operators are not numbers. 1,2,3,4,5… are numbers, whereas '+', '-', '•', or ':' are binary operators. They are binary because they combine two numbers (called 'operands') to obtain another number (for example, 1+2=3). Therefore, it makes no sense to state that if $O|\Psi\rangle = \lambda|\Psi\rangle$, then $O = \lambda$, because you are then equating an operator with a number—that is, mistaking apples for bananas. Moreover, the operators used in QM operate on functions rather than on numbers. Perhaps the simplest non-trivial unary operator is the derivation operator $\frac{d}{dx}$. It is unary because it operates on one function to obtain another function. For those who know how to take derivatives (and for those who do not, don't worry, as we will not need them anymore), an example could be $\frac{d}{dx} e^{2x} = 2e^{2x}$ (and nobody would equate $\frac{d}{dx} = 2!$). Quantum operators—that is, observables—are also called 'Q-numbers' because they have a formal analogy with numbers. However, remember, they aren't numbers at all.

Once that has become clear, here are some examples of observables. The simplest one is the position operator:

$$O_x = X = x.$$

If we want to know the position of a particle, we multiply the wavefunction with a position variable x and then take the modulus square to obtain, as usual, the probability density in function of x. This tells us something about its whereabouts.

The momentum operator, that observable which tells us something about the momentum of the particle, has the form of a derivative operator. (For those who struggle with the notion of a derivative of a function, see Appendix A Ie.):

$$O_P = P = -i\hbar\frac{d}{dx}.$$

Taking the derivative in space of the wavefunction (times the Planck constant bar and an imaginary number) will furnish a mathematical expression that tells us something about the speed of the particle that the wavefunction describes.

Another important quantity in QM is the 'expectation value' of an operator – that is, the average value one obtains by measuring an observable many times. It is operatively a fundamental notion in QM because in the laboratory, what one usually must tackle in the real world are averaged measured values. For example, for the position operator $X = x$ of a particle, one integrates its position x, weighting it by the probability of finding it in x or, in analytical terms (recall *Eq. 9* and Eq. 10):

$$\langle x \rangle_\Psi = \int_{-\infty}^{+\infty} \psi(x)\, x\, \psi(x)^* \, dx.$$

It is the average value of the position x of a particle that one expects to obtain after a large number of measurements. More generally, this can be written in a more compact form by using Dirac's notation, stating that the expectation value of an operator A in state Ψ is defined as:

$$\langle A \rangle_\Psi = \langle \Psi | A | \Psi \rangle, \qquad Eq.\ 15$$

with right-hand side Dirac bra-kets as defined in the integral above.

Another operator of paramount importance in QM is the energy operator, the 'Hamiltonian', which has a bit more complex structure. It is given by a second derivative in space (times $-\frac{\hbar^2}{2m}$, where m is the mass of the particle) plus a function $V(x)$, which represents the particle's potential energy. By '*potential energy*', physicists mean the energy content of an object that can arise due to its position inside a force field (such as an object in a gravitational field or an electric charge in an electric force field) or stresses within itself (for example, an expanded spring) or in the form of chemical energy (the chemical bonds between atoms and molecules), etc. Without us getting into additional rigorous definitions and details, just imagine it as a sort of 'stored energy' (here, $V(X)$ can be the function which tells us the stored energy an electron has at a certain distance x from the nucleus), which, however, can be transformed into another form of energy for practical purposes. The archetypal example is the transformation of potential gravitational energy (say, the potential energy of a stone at the top of a hill) into kinetic energy (the stone rolling down the hill and acquiring a certain speed). The Hamiltonian operator is written as follows:

$$O_E = H = -\frac{\hbar^2}{2m}\frac{\partial^2}{\partial x^2} + V(x)$$

It is this latter observable in particular which leads to the famous '*time-independent Schrödinger equation*':

$$H|\Psi\rangle = E|\Psi\rangle .$$

The Schrödinger equation is nothing other than the eigenequation for a system's energy states. Once the wavefunction, or the state vector describing the system, is given, one applies the Hamiltonian as prescribed by the eigenequation. We will obtain as eigenvalue E the energy of the system.

This is, however, not the most general equation because, as the name indicates, it is time-independent, while physical systems are, in many cases, time-dependent. Therefore, the time-dependent energy operator is called '*evolution operator*' and is given by:

$$O_E = H(t) = i\hbar\frac{d}{dt}.$$

That's why you will frequently find the *'time-dependent Schrödinger equation'* in this form:

$$H(t)|\Psi\rangle = i\hbar\frac{d}{dt}|\Psi\rangle.$$

The Hamiltonian operator also defines the *'time evolution'* of the wavefunction from a time 0 to a time t. In particular, if H is independent of time, given a state $|\Psi(0)\rangle$ at an initial time t=0, the state of the quantum system $|\Psi(t)\rangle$ at any subsequent time t is *'generated'* by the so-called *'unitary operator'* $U = e^{-iHt/\hbar}$ as:

$$|\Psi(t)\rangle = U\,|\Psi(0)\rangle. \quad \textit{Eq. 16}$$

Its main physical interpretation is that it is an operator that does not describe an observable (that is, a measurement), but the evolution in time of a quantum system. It changes the phase of the wavefunction or state vector but not its length. This 'length conservation' reflects nothing else than the necessity to conserve the sum of all the probabilities associated to the eigenstates before and after the state evolution, that is, according to Eq. 14 it must hold:

$$|\langle\Psi(0)|\Psi(0)\rangle|^2 = |\langle\Psi(t)|\Psi(t)\rangle|^2 = 1.$$

Therefore, its name 'unitary operator'. Moreover, note that if U is the unitary operator that evolves the quantum state forward in time, its complex conjugate $U^* = e^{+iHt/\hbar}$ sends it backwards in time by an interval t. Then it becomes obvious that applying first U^* and U consecutively it preserves the state, that is, an equivalent formal definition for a unitary operator is that it satisfies the general relation:

$$U^*U = 1, \quad \textit{Eq. 17}$$

With one being a scalar (just the number '1') or, more generally, an identity operator.

More on this later, as we will see this operator reappear throughout our journey and discuss its properties in different contexts again.

The Schrödinger equation is important because it helps us obtain the energy states of any quantum system, such as the energy levels of atoms and their nuclei. In fact, the Schrödinger eigenequation allows for the calculation of the energy spectra of atoms. The potential energy V(x) of the electrons around the nucleus depends on the number of protons in the nucleus. If one sets particular boundary conditions on the structure of this potential (for instance, that it is spherical and that it decreases according to an inverse law, etc.) and the wavefunction must be continuous, the Schrödinger equation becomes a solvable (second order) differential equation whose solution will furnish the energy for each of these states and its probability amplitude $\Psi(x)$ to find the particle somewhere around the atomic nucleus.

For completeness, since we will make use of it when dealing with quantum decoherence, let us point out how, once the Schrödinger equation has been solved and the eigenstates with its eigen-energies are known, it is possible to calculate the '*transition amplitudes*' in accordance with Eq. 13, that is, the probability that an atom transitions, say because of the scattering or absorption of a photon, from an initial state $|\Psi\rangle$ to a final state $|\Phi\rangle$. Dirac's notation allows to determine the probability that this transition will take place as:

$$Prob_{\Psi \to \Phi} = |\langle \Psi | \Phi \rangle|^2 . \quad Eq.\ 18$$

For example, the probability that an atom in isolation spontaneously transitions from an energetic lower eigenstate $|E_0\rangle$ to a higher one $|E_1\rangle$ is zero, which implies that $\langle E_0 | E_1 \rangle = 0$. In this case one speaks of '*orthogonal state vectors*'. Geometrically this is displayed in *Fig. 62* as being the eigenstate vectors perpendicular to each other. However, for any quantum system which is subjected to an outside interaction, such as in a scattering or a measurement process, the state vectors may no longer be orthogonal and Eq. 18 acquires a non-zero value. There is then a certain chance that due to this interaction a state transition occurs (in case of atoms, eventually by emission or absorption of photons).

At any rate, we always have to keep in mind that probability amplitudes are only a probabilistic description of the quantum system. The eigenequation formalism, with its operators, eigenvectors, and eigenvalues, is a mathematical description of the probability of finding a particle in a specific position, with some specific momentum and some (discreet or continuous) energy state. There is no such thing, as in the Bohr's atom, as particles flying around a nucleus. There are no particles at all but only the probability density function wrapped in some form or another around the center, the nucleus of an atom.

In the case of the simplest possible atom, only with an electron immersed in the potential field of a proton, the hydrogen atom, will one obtain a wavefunction which describes the simplest probability density, that of the hydrogen atom energy '*ground state*': just a spherical probability density centered around one proton, as shown in Fig. 64.

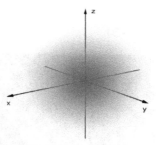

Fig. 64 The orbital of the hydrogen atom in its ground state.

No nucleus, no electron is visible as a point particle. What this represents is the probability density for finding the hydrogen atom's electron. It is just a graphical – but quite effective – representation, the *'atomic orbital'*. Orbitals are the squared modulus of the wavefunctions obtained from Schrödinger's equation, the graphical representation of which highlights the statistical probabilistic nature of the quantum world. The more intensely shaded regions represent a higher probability of finding the electron in some place, while the less intense regions represent a low probability density. We get some sort of cloudy and fuzzy sphere, the *'electron cloud'* or *'electron configuration'*, which no longer has anything left of the clear, well-defined, deterministic, and precise representation of CM. However, Fig. 64 depicts the hydrogen atom only in its ground state. As we know, atoms are capable of absorbing, in a quantized manner, quanta of energy and acquire other energy discrete levels. If an atom absorbs a photon, the electron cloud will acquire a different configuration – that is, a different wavefunction with another probability distribution which will describe the excited atom.

Fig. 65 shows several examples of such possible configurations. These are 2D slices of the 3D probability distributions. It is what we get when we cut the atom along a slice that goes through its central nucleus, so to speak. Each state defines the form and structure of the orbital and the slice. They are 3D standing waves, the kind of 3D analogue of the 1D standing waves in a box we already considered to explain the blackbody radiation (see Fig. 11). The difference, however, is that an atom does not have a solid confining boundary, as do the walls of a box, but is smoothly delimited by its nuclear electric potential.

The atomic electron configurations can be described using three atomic quantum numbers. The first is the *'principal quantum number'*, labeled n, which represents the atom's energy level. We have already encountered this in the Bohr atom model with circular electronic orbits. The second is the *'angular'* or *'azimuthal quantum number'*, labeled l, which characterizes the total angular momentum of the electron and specifies the shape of the orbital (in that specific energy level n).

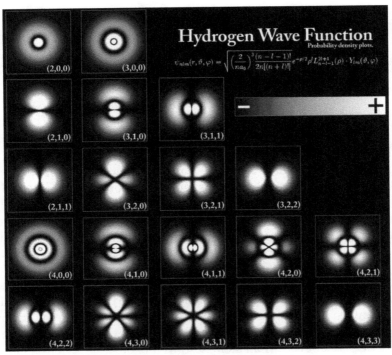

Fig. 65 Some orbitals of the hydrogen atom.

This is the additional quantum number that emerges when one considers the Sommerfeld extension from circular to elliptical electronic orbits. The third quantum number, the so-called '*magnetic moment quantum number*', labeled m, describes the orientation (of the orbital in energy state n and angular momentum l) along a specific direction.

From the mathematical quantization procedure we have described above, and by applying a formal description no longer in Cartesian coordinates but in spherical coordinates, it is possible to show how all these three numbers arise naturally and describe the different states of an atom. A nice probability density cloud is associated with every combination of the quantum numbers. Note also how quantum states exist which have the same energy but are represented by different orbitals – that is, by different wavefunctions and probability distributions in space. These are called '*degenerate quantum states*'.

Yet apart from these technicalities, which we mentioned out of completeness and which would require a much more rigorous and lengthy mathematical-physical description that goes beyond the aims of this book, what you must keep in mind is that what you are seeing here are probability distributions (another decisive difference from the blackbody cavity, where the standing waves considered were EM waves). This means that if we would try to detect the position of an electron inside these orbitals, the wavefunction would collapse instantly upon the act of measuring a point particle. However, before that, in QT, we must conceive of the system not in a reductionist sense as made of particles but as a single, whole entity described by a fuzzy electron cloud represented by an orbital. It is like the case of the double slit experiment: Before the photon hits the screen, we must conceive of a wave describing the photon. At the instant of interaction, it collapses to a point on the screen. Here, we have the same situation: Before any attempt to measure something, it makes no sense to say that the electron has a specific position around the nucleus of an atom, but only that we have a certain probability of finding it there, highlighting the purely wavy and probabilistic nature of the quantum world.

These sorts of calculations, which all rely on the aforementioned mathematical formalism, in particular on Schrödinger's equation, can, of course, be applied to much more complex atoms than the hydrogen case. They can also be applied to atomic nuclei and molecules. In fact, it is in this way that nuclear and atomic physics have been successfully understood and exploited, even today. Moreover, molecules and several properties of matter can be explained. Solid state physics is one of the most prolific sciences of modern times. Schrödinger's equation – and much more in general, the eigenvalue equations of Dirac's notation, which builds upon the Born rule – turned out to be a powerful mathematical tool which led to the quantum revolution of the 20th century. This allowed physicists to develop a sophisticated but consistent theory that proved to be correct over and over again, providing a formidable theoretical background that allowed applied physics to build the material basis upon which the modern electronic devices industry was revolutionized. Finally, on the theoretical side, adding SR to this non-relativistic QM led to the standard model of particle physics, which can be considered one of the most successful and precise scientific theories in the history of science.

III.   The quantum world of spinning particles

1. Angular momentum in classical mechanics

Before focusing on the mysteries and paradoxes of the quantum world, we must learn a couple of elementary concepts – such as angular momentum and spin – that we will find throughout the rest of this book.

So, what is angular momentum? In CP it is not very hard to understand it. It is the amount of rotational momentum, the quantity of motion of rotating bodies. Therefore, let us first look at the notion of the angular momentum of a single particle rotating around a center, as shown diagrammatically with vectors in Fig. 66.

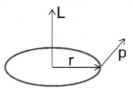

Fig. 66 The angular momentum of a particle.

Recall how the linear momentum of a particle which moves unperturbed along a straight path is proportional to the product of its mass m times velocity v (at least in non-relativistic CM): $p = m \cdot v$. Similarly, the general expression of the angular momentum L of a particle with rotational velocity v about an origin is defined as:

$$L = r \times p = r \times m \cdot v, \qquad \text{Eq. 19}$$

where r is the position vector of the particle relative to the rotational origin, and the cross denotes the vector or cross-product (a sort of multiplication for vectors). Please note that both the linear and the angular momentum are vector quantities, which means they have a magnitude and direction. Letters representing vectors are written with an arrow or in bold. (Here, we use the latter.)

In the event of a body made of n particles rotating around a center external to itself (that is, orbiting around a center of origin), the '*orbital angular momentum*', L_{orb}, can be obtained by determining the sum of the linear momenta of each of its constituent particles – the sum over all the particles of masses m_i times the respective rotational velocity v_i for each, as:

$$L_{Orb} = \sum_i^n r_i \times m_i \cdot v_i \, ,$$

again with r_i as the i-th particle's distance from that axis and n as the number of particles.

However, an extended body can also rotate around itself. A typical example is that of the Earth, which not only orbits around the Sun but (as everybody knows) spins once around its polar axis approximately every 24 hours.

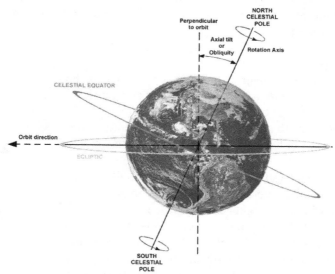

Fig. 67 The Earth's rotation.

In this case, for bodies whose interiors have a complex mass distribution, like the Earth (the density distribution inside the earth can be quite complicated), one must calculate the so-called *'moment of inertia'* of a body.

Inertia is a physical quantity that measures the resistance a material body exerts upon the change of motion. In the case of a point-like particle, it is simply its mass. In the case of a body with an extension and a more or less irregular complex geometry and internal matter distribution, it is a scalar quantity, a number *I*, which can be calculated using mathematical procedures.

In general, the *'spinning angular momentum'* L_S is given by:

$$L_S = I \cdot \omega, \quad Eq. \ 20$$

where ω is the mathematical expression of the *'angular velocity'* which is the measure of how fast it spins around itself – that is, how quickly it completes a 360° rotation per unit time, which, in the case of a spherical body, is just $\omega = \frac{v}{r}$, the rotational speed of the body divided by its diameter.

Therefore, we must distinguish between the angular momentum of a particle or a body moving around a center (e.g. a planet around the Sun) and the angular momentum of a body around its central axis (e.g. the above-mentioned case of the angular momentum of the Earth spinning around its polar axis). These are two closely related but slightly different quantities. One is an orbital angular momentum, while the other is a spinning angular momentum, also known simply as '*spin*'. The total angular momentum is the sum of the two:

$$L_{Tot} = L_{Orb} + L_S \, .$$

An important universal law to keep in mind is that of the '*conservation of angular momentum*', which is a direct consequence of the conservation of energy (i.e. energy can never be created or destroyed, but only transformed). This law states that the total angular momentum cannot change – that is, it is conserved unless an external torque acts on the rotating body.

In Eq. 19 the radius r or the speed v are allowed to change, but the overall momentum L must remain constant. If a particle gets closer to the rotation center (r decreases), the angular velocity must increase to maintain L constant. This also holds for any extended body. A common example illustrating this principle is that of the ice-skater. When ice-skaters bring their arms closer to their bodies, the angular velocity increases, and vice versa. This act decreases the inertia I; according to the law of conservation of angular momentum, an increase of ω must follow to maintain L_S constant (see Eq. 20), and vice versa.

Fig. 68 Conservation of angular momentum: Example of the ice-skater.

Now that we have been introduced to the main concept and principle of the angular momentum of CM, let us apply this to QM.

2. Spin, the Stern-Gerlach experiment and the commutation relations

In QM, one is typically more concerned with single particles or many particle systems than with extended bodies. So, what about elementary particles, like electrons? It is not difficult to extend the notion of the orbital angular momentum of a material particle having a mass like, for example, the electron in the case we conceived of it in Bohr's atomic model, as a particle flying around the atom's nucleus. This conception is dubious enough because we should actually have in mind the atomic orbital model, not Bohr's atomic model. In any event, so far this picture of reality still works: The orbital angular momentum of the electron can be defined as the product of its mass times the velocity at which it orbits the nucleus times the orbital distance from it (Eq. 19). However, this is where the analogies to the classical world end.

In fact, what about the spin – that is, what in QM is called the '*intrinsic angular momentum*' – of a particle rotating around an axis that goes through its center, as in the case of the earth? The problem is, this analogy breaks down because we think of so-called 'elementary' particles as a mathematical point-like object. Thus, it becomes somewhat unclear what the spin of a point structure should mean at all. What is a point that spins around itself? Intuitively and even mathematically, it doesn't make much sense.

So, either electrons are not point-like, but no experiment has so far revealed any internal structure (at least not as of 2019), or we must consider the notion of the spin of particles in QM as something that must not be viewed as we do intuitively at the macroscopic scale. For this reason, in QM, one more loosely speaks of particles 'carrying' an intrinsic angular momentum, or spin, without suggesting literally the image of a spinning sphere, as in classical theory.

And yet, we know that even elementary particles such as electrons, protons, and neutrons have a tiny but measurable spin. Can the spin of a particle as tiny as an electron be measured? The answer is positive, and even surprising. It is possible to measure and associate a spin with elementary particles. The interesting point is that, as we saw energy being quantized in QM, the spin of particles also turns out to be quantized. Nature allows only discrete – not continuous – quantities of intrinsic angular momentum! Spin in QM is a quantized dynamical property of all particles.

How physicists reached this conclusion will be elucidated next with the Stern-Gerlach experiment. It is easier to understand how this works if we first introduce some basics and anticipate some of the results.

For elementary particles, the spin has only two possible values. If we take as a convention the vertical axis (see Fig. 69) being the upward z-axis, we can show experimentally that, say an electron or a photon, has only two possible spins along that axis: a fixed specific amount of spin-up or spin-down states. We will never, ever observe the electron having other amounts of spin and being directed towards another direction in between. Or, if you would like to stick to the classical intuitive understanding, elementary particles rotate clockwise or counter-clockwise around an axis, always with the same angular frequency.

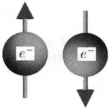

Fig. 69 A spin-up and spin-down electron.

What must be added to this picture is the fact that we should not forget that several particles, like electrons and protons, are electrically charged particles. (Neutrons are not.) They all possess an equal tiny charge, which is also taken as the elementary electric charge, labeled 'e', which amounts to $e = 1.672 \times 10^{-19}$ C, where [C] stands for '*Coulomb*' and is the standard international unit for the electric charge. This is an extremely small charge. For example, a typical household electric device (say, a small lamp) that works with an electric current of 1A ([A] stands for '*Ampere*', the unit of the intensity of an electric current), has something of the order of 6×10^{18} (six quintillion, that is, six billions of billions) electrons per second flowing through its circuitry.

However, despite its small value, the electric charge of an electron can be measured using relatively simple considerations and experimental arrangements. In fact, electrically charged particles produce an electric field in their surrounding space, which interacts with other charged particles or with magnetic fields. They are also themselves the source of magnetic fields. Every moving electrically charged particle always produces a corresponding magnetic field. A cable through which an electric current – that is, electrons – flows will also manifest a magnetic field. This is always true and is a fundamental law of physics: Wherever electric particles travel throughout space or through a conductor, they will always produce a magnetic field. (The construction of electromagnets relies on this principle.) A loop of electric current, a bar magnet, an electron, a molecule, and a planet all have magnetic moments, as they all contain, in one form or another, circulating electric charges. This is something we imagine also happening with the Earth's magnetic field. The Earth's interior must contain a huge amount of magma, of hot currents of a magma fluid, which is electrically charged. Its movement around the Earth's axis causes the Earth's magnetic field to build up. The opposite is true as well. Wherever a magnetic field changes in time, an electric field will appear. This is what we already hinted at when discussing the nature of light as an oscillating EM field and it is at the foundation of Maxwell's equations.

Therefore, an electrically charged spinning object is also expected to display some magnetic field, as it is a charge in rotational movement. A particle's magnetic field is induced by the charge flow around itself due to its spin. This makes every elementary particle, like electrons and protons, also tiny magnets. Therefore, not only do they possess an electric charge but because they possess a spin, they must produce a tiny magnetic field as well. And because only two fixed spin states are possible, the result is a '*magnetic moment*', with the magnetic field lines directed in one or the other direction according to the particle's spin orientation, as shown in Fig. 70.

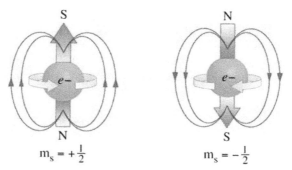

Fig. 70 Spin orientation and magnetic field lines of an electron.

The magnetic moment of a magnet is a physical quantity that tells us how strong the magnet is. More precisely, it is a vector quantity which determines the torque that the magnet (here, the particle) will experience in the presence of another external magnetic field (say, some produced in the laboratory with electromagnets).

And, if we can measure a particle's magnetic moment, we can be sure that it possesses a spin and we can evaluate its magnitude. Of course, because the spins of quantum objects are quantized, the same must hold for its magnetic moment. Electrons and protons, therefore, have only two magnetic moments due to their corresponding spin, $S = \pm \frac{1}{2}$, which are accordingly labeled as $m_s = \pm \frac{1}{2}$.

This is precisely what, in 1922, the German physicists, Otto Stern (1888-1969) and Walther Gerlach (1889-1979), were able to demonstrate.

Fig. 71 Otto Stern (1888-1969).   *Walther Gerlach (1889-1979).*

In particular, they showed that not only particles like the electron but also atoms have quantized magnetic and intrinsic angular momentum. The Stern-Gerlach (SG) experimental setup was constructed as follows. They sent a beam of silver atoms, previously prepared in a source of hot atoms (in that case, a furnace) and collimated it by a slit, sending it through an inhomogeneous magnetic field created by magnets that had a special form and orientation, of the sort seen in *Fig. 72*.

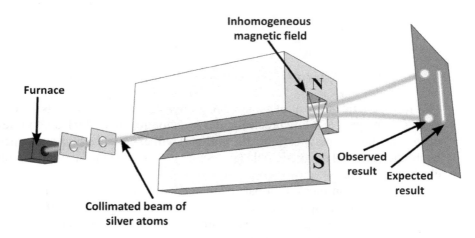

Fig. 72 The experimental setup of the Stern-Gerlach experiment.

The choice of silver atoms was due only to practical considerations. Silver atoms are easier to handle and have only one unpaired electron in their outer atomic shell, while the electric and magnetic properties due to all the other inner electrons are cancelled mutually (we will explain this further when we are dealing with the Pauli exclusion principle) and electrically shielded. Therefore, the measurement of the magnetic moment of a silver atom reduces to that of one single outer shell electron.

Now, when an atom (or a particle, or whatever object has a magnetic moment) is sent through an inhomogeneous magnetic field, it experiences a force and therefore is deviated in one or the other direction relative to the poles of the magnet that is relative to the direction of the magnetic lines and according to the orientation of its own magnetic moment.

It is quite intuitive to understand why this must be so: Just as a couple of magnets must repel or attract each other, the SG magnet and the particles or atoms, which are also tiny magnets, must mutually repel or attract each other according to the directions of their magnetic polarities. That is, if one sends the particle which has a spin-up, say, per convention relative to the z-axis, through the SG-device with its North-South magnetic poles directed along the z-axis, it will be deviated upwards. On the other hand, the spin-down particle will be deviated downwards. And if the magnetic moment of quantum particles is indeed quantized in only two possible states, we will observe only an up or down deviation in the SG experiment, which must therefore create two distinct spots on a detecting screen that the particles strikes.

Otherwise, if the magnetic moment were continuous, that is, would not be a discrete physical property of particles, the deviation of the SG magnets induced on them would also be continuous, producing a continuous line on the detection screen, as expected by CP. This was a conceptually simple method of showing that atoms and particles indeed possess a quantized intrinsic angular momentum. Stern and Gerlach observed, always and only, the binary outcome, not the expected classical result. What they showed with this is that the quantization of spin is not simply speculation but established experimental fact.

Before we discuss other interesting discoveries one can make with a SG-device, let us fix this first result with the Dirac notation we discussed in the lecture on the state vectors. The two possible spin states could be labeled with the $|+\rangle$ ('ket-plus') and $|-\rangle$ ('ket-minus') symbols. If not further specified, it is convention to assume that these states are measured against whatever axis or the standard z-axis. Other symbolic conventions are to also label them as $|+\rangle_z$ and $|-\rangle_z$ or $|S_z\rangle_+$ and $|S_z\rangle_-$ or $|S_z; +\rangle$ and $|S_z; -\rangle$, and so on, especially when it is important to distinguish them from the other x and y directions. We will use the first ones. However, because no universal standard convention exists in this regard, please learn to be flexible enough to read them interchangeably according to the context!

The important point to keep in mind is that SG-experiments can easily show that the quantized character of the spin does not change even with a change in the direction along which the measurement is taken. Whatever axis you choose in the 3D space, you will always get, as a result of the measurement, a spin-up or spin-down, a $|+\rangle$ or $|-\rangle$ particle. As strange as that might seem, the fact is, in the quantum world there is no longer such a thing that can smoothly vary its spinning direction and rotational speed, as we imagine it, for example, with a spinning billiard ball. Instead, we have quantum objects which, once a direction has been chosen, can acquire only two possible spin states along that direction – nothing in between.

The amount of intrinsic angular momentum can also be measured. One can relate it to the separation of the two spots on the SG-screen; it turns out to be equal to $\pm\frac{\hbar}{2}$ ($\hbar$ being, as usual, Planck's constant divided by 2π). In fact, Planck's constant has the dimension of an angular momentum and, if we consider that it is a value of the order 10^{-34} Js, when we recall what we have said in the section on the blackbody radiation, we realize that it is an incredibly small but not zero amount of angular momentum.

We have also seen how, if a measurement on a particle takes place, this is represented in mathematical abstract terms by operators that we called observables, acting in a Hilbert space on vector states, that is, the kets in the Dirac notation. On that occasion, we described the measurement process by means of an operator eigen-equation with an eigenvalue representing the measured quantity. Because the spin is a physically measurable quantity, it must have an associated operator: the spin operator S. A spin operator is also defined along one of the three dimensions of space, again, usually that directed along the z-axis as S_z. However, there is no reason to restrict ourselves to the measurement along one axis. Three possible spinning directions exist in a 3D space. We can also consider measuring the spin of an object along the other two x- or y- spinning axes with operators S_x and S_y respectively.

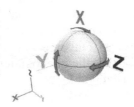

Recall that in QM, every act of measurement represented by an operator – the observable of the physical property measured – acts on a wavefunction or state vector and delivers the eigenvalue – the quantity one measures – through the eigenvalue equation. If the particle is in an eigenstate (the same state which will always turn out to be, even after repeated measurements), the spin operators S acting on the eigenvector of an electron in a spin-up or spin-down eigenstate $|\pm\rangle$ (the spin measurements of the SG-device) for all three axes is represented:

$$S_z\,|\pm\rangle_z = \pm\frac{\hbar}{2}|\pm\rangle_z\,;$$
$$S_x\,|\pm\rangle_x = \pm\frac{\hbar}{2}|\pm\rangle_x\,;$$
$$S_y\,|\pm\rangle_y = \pm\frac{\hbar}{2}|\pm\rangle_y\,.$$

The measurement process along the z-axis is written as S_z acting on the eigenvector $|+\rangle_z$ and resulting in a $+\frac{\hbar}{2}$ eigenvalue, the result of the measurement, times the $|+\rangle_z$ eigenvector itself. The same holds for the particle in spin-down state, but with the opposite sign, and so on.

However, generally speaking, the outcome is not known and the probability of finding the particle in one or the other state is 50%. We have learned that the probability of finding the system in a specific eigenstate is encoded in the coefficients of the state vector which contains all the possible states in which a system can be. Therefore, the spin state of an electron is written as follows (omitting, for sake of simplicity, the axes' labels):

$$|\Psi\rangle = \frac{1}{\sqrt{2}}|+\rangle + \frac{1}{\sqrt{2}}|-\rangle. \qquad Eq.\ 21$$

Let us closely inspect this state vector equation. The ket $|\Psi\rangle$ is *not* (or not necessarily) the state in which the particle is observed to be. It is the state before the measurement and comprises all possible outcomes. We might say that it describes a potentiality of realization. Because, for an elementary particle like an electron, only two possible measurement outcomes exist for the spin states (namely, the $|+\rangle$ or $|-\rangle$ eigenstates) and because there is no reason to believe that Nature prefers one over the other, they must both have an equal 50% probability of being measured (an easy fact to determine measuring how intense the spots on the SG-screens are). This probability is encoded in the coefficients $\frac{1}{\sqrt{2}}$. In fact, recall that the squared modulus of these coefficients furnishes us with the probability of observing one or the other associated eigenstates realized by the measurement process. Here, $\left|\frac{1}{\sqrt{2}}\right|^2 = \frac{1}{2} = 0.5$, which means 50% each.

Eq. 21 is one of the most typical equations you will find in most treatises on QP and will follow us throughout this journey in one form or another. Notice that the state vector $|\Psi\rangle$ represents the state of the system *before* the measurement as the sum of mutually exclusive possible outcomes, the eigenstates $|+\rangle$ and $|-\rangle$. Meanwhile, these eigenstates are the states in which the system will be found *after* performing the measurement which will 'project' our measurement apparatus to a macroscopic state that displays a spin-up or a spin-down state (for example, with a pointer), but not both.

With these fundamentals in mind, we can discuss other revealing experiments that one can perform with SG-devices.

Fig. 73 depicts, in the form of a block diagram, the SG-device for the two possible measurement outcomes.

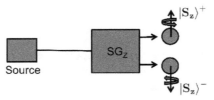

Fig. 73 Block diagrams of the two possible outcomes in Stern-Gerlach experiments.

The source sends spin $\pm\frac{1}{2}$ particles, say, silver atoms or electrons, towards an SG_z device which measures the spin along the z-axis. If the electron is in an $|+\rangle_z$ or $|-\rangle_z$ spin eigenstate, it will be deviated towards the upper or lower part of the apparatus, respectively.

Let us begin with the SG_z measurement outcome of the spin-up electron and construct, in different steps, the experimental procedure. Begin with the diagram in Fig. 74. Physicists speak of a system (here, just a single particle) 'prepared' in the $|+\rangle_z$ eigenstate.

However, we are good experimentalists and want to be absolutely sure that when the electron went through the SG_z device, its state was not destroyed by the measurement itself and that it is still in the $|+\rangle_z$ eigenstate.

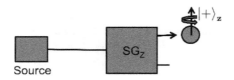

Fig. 74 Measurement (A): The SG_z -device 'prepares' an electron into a $|+\rangle_z$ eigenstate.

To do that, we simply cascade another SG_z device, as in measurement (B) of Fig. 75.

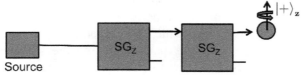

Fig. 75 Measurement (B): Two identical SG_z -devices are cascaded to confirm measurement (A).

If we made no mistakes and our laboratory equipment worked fine, one would again observe the electron coming out of the corresponding $|+\rangle_z$ output. This is an experimental fact. So, we are sure that the measurement as such does not destroy the electron's intrinsic angular momentum along the z-axis. We feel authorized to say that, in fact, the electron is in that particular spin-up eigenstate.

However, things are not as easy as that. To demonstrate this, let us extend our measurements along other directions. Consider measurement (C) as in Fig. 76.

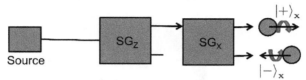

Fig. 76 Measurement (C): Two SG-devices:
One selects the z-spin particles, the other measures their x-spin.

Here, we first select, again, the electron with the $|+\rangle_z$ spin with the first SG$_z$-device, which measures the spin along the z-axis, and then send it through an SG$_x$-device, one that measures the spin along the x-axis. (To do that, one has only to appropriately rotate the SG-device's magnets). To ensure that the system is functioning properly, we may also send the particle through more than one SG$_z$-device, just to be sure it is in the $|+\rangle_z$ eigenstate, and then send it into the second SG$_x$-device. What will be observed as output from the latter is that both spin-x eigenstates states will show up. That is, if you send several electrons through such a system of cascaded SG-devices, you will have a 50% chance of seeing it come out in a spin-up and a 50% chance of seeing it in a spin-down eigenstate, always. This is something we can measure empirically with experiments.

Initially, you might have no problem with that. We may simply believe that of all the particles initially prepared in a state $|+\rangle_z$, 50% have a $|+\rangle_x$ spin while the other 50% are in a $|-\rangle_x$ spin. After all, we are dealing with two different spin directions that we think of as being independent from each other. For our conventional classical logic, it sounds entirely legitimate to state that, in preparing the system in the first stage in a $|+\rangle_z$ state without caring in what state it is along the x-axis, it makes sense that statistically we obtain 50% of $|+\rangle_x$, and 50% of $|-\rangle_x$ particles at the end of the chain. And, if our SG-devices are ideal and do not destroy the eigenstates during any measurement process, we might effectively consider the electron that spins along the z-axis to really be in a spin-up state. There is no reason to believe otherwise.

This interpretation, however, will not hold up because the experimental evidence will show us otherwise. In fact, we are scientists who want to be very cautious and maintain absolute certainty that the intrinsic angular moment along the z-axis is still there.

Therefore, we again measure the spin along the z-axis of one of the outcoming particles of measurement (C) – say, that with spin-up in the $|+\rangle_x$ eigenstate – and we cascade another SG$_z$-device which measures the spin along the z-direction again, as shown in *Fig. 77* of measurement (D). What we will discover is, now that the particle's spin-up eigenstate along the z-axis is gone, it is no longer determinate and becomes uncertain, as we will again have a 50% chance of getting the spin-up or spin-down eigenstate again.

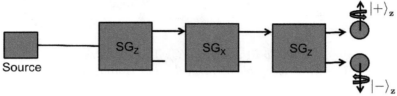

Fig. 77 Measurement (D):
Three cascaded SG-devices: Spin measurement z-x-z.

If you are not surprised about that, almost certainly you did not get the point, so please read it again. For our classical understanding of the world, this is a complete contradiction. In measurement (B), didn't we twice measure the z-spin to ensure that the measurement did not destroy a particle's intrinsic angular momentum? We can repeat the same experiment of measurement (B) with the SG$_x$ device cascading another SG$_x$ and we won't see any destructive activity along the x-axis either. If applied twice, it won't disturb the system. And yet, if we first measure the spin along the z-axis and then along the x-axis, we see that Nature does not allow for that. If one repeats the same experiments along the y-axis in all its possible combinations, the result will not change.

What all this tells us is that we cannot know, at the same time, the value of the spin along two axes. And again: This is independent of any interaction between the measuring apparatus and the particle. It is important to keep in mind that this has nothing to do with the interaction between the observing device and the observed object (and even less with the supposed mind or consciousness of a human observer).

Does this remind you of something? It should. Here, we have, again, a phenomenon which is strongly reminiscent of Heisenberg's uncertainty principle. In that case, the position and momentum were indeterminate; they could not be measured at the same time. Here, they are the spins. In the case of Heisenberg's uncertainty principle, this is expressed with position and momentum operators. Here, the spin operators along three directions come into play. However, the underlying principle is of the very same nature. The precise knowledge of one observable seems to be incompatible with certain knowledge of another observable. Once you measure one dynamic variable, you are no longer allowed to extrapolate anything about the state of the previous one. In general, once a measurement prepares a quantum system in a specific eigenstate, that is, it projects the state vector $|\Psi\rangle$ onto one of its eigenbasis vectors $|\pm\rangle_x$, $|\pm\rangle_y$ or $|\pm\rangle_z$, it also renders indeterminate the outcome associated with the other two eigenstates.

This automatically implies that the order in which the measurement is performed is important. Performing a measurement on a quantum system first with the observable S_x and then with S_z will not lead to the same result as that stemming from performing it first with the observable S_z and then S_x. This is because, in the former case the system is projected, say, in eigenstate $|+\rangle_x$ with $|\pm\rangle_z$ and $|\pm\rangle_y$ uncertain, while in the latter case the system is projected, say, in eigenstate $|+\rangle_z$ with nothing that can be said about the $|\pm\rangle_x$ and $|\pm\rangle_y$ eigenstates. In symbols:

$$S_x S_z |\Psi\rangle \neq S_z S_x |\Psi\rangle,$$

or also

$$(S_x S_z - S_z S_x)|\Psi\rangle \neq 0.$$

The same applies to the other spin operator combinations $S_x S_y$ and $S_y S_z$. Physicists and mathematicians say that the '*observables do not commute*' and express this formally with the so-called '*commutation relations*' written with square brackets, the '*commutators*'. For two generic operators $\hat{A}$ and $\hat{B}$ (the caps $\frown$ are symbols to distinguish operators from numbers, that is, scalar or vector quantities), the commutator is defined as: $[\hat{A}, \hat{B}] = \hat{A}\hat{B} - \hat{B}\hat{A}$ (for example, $[S_x, S_z] = S_x S_z - S_z S_x$ and so on). So, when observables do not commute, this simply means that we can't know them both at the same time.

While we do not do this here because it would require a formal algebra that goes beyond the scope of this book, it can be shown that the spin commutators obey the following rules:

$$[S_x, S_z] = i\hbar S_y,$$
$$[S_y, S_z] = i\hbar S_x, \qquad Eq.\ 22$$
$$[S_x, S_y] = i\hbar S_z,$$

with i the imaginary number and, as usual, $\hbar$ as Planck's constant divided by 2π.

This allows us to formally formulate Heisenberg's uncertainty principle. In the case of the uncertainty principle, the position and momentum operators do not commute because, as we know, only one or the other observable – not both – can be determined with high precision. One can show that, if one applies the position and momentum operators consecutively on the wavefunction, one will not obtain a zero result but, again, a quantity which is the imaginary number times the Planck constant. Therefore, formally, Heisenberg's uncertainty principle is stated as:

$$[x, p_x] = i\hbar,$$
$$[y, p_y] = i\hbar, \qquad \text{Eq. 23}$$
$$[z, p_z] = i\hbar,$$

where the operators have been generalized to a 3D position (x, y, z) and momentum space (p_x, p_y, p_z).

It is important to keep in mind the meaning of the commutation relations and what physicists mean when they speak about observables that do not commute. The non-zero commutation relations expressing the impossibility of establishing the value of two observables at the same time is one of the most fundamental principles of QM. We can already anticipate that this is the formal mathematical expression of Bohr's '*complementarity principle*': When observables exist that we cannot know precisely at the same time, they are 'complementary' to each other.

There are, however, observables that commute – that is, the commutation relations result is zero. This is the case when one measurement does not render the previous one uncertain. This is always the case with one observable with itself, which is to say the same measure is made twice. For instance, what we saw in case (B): $[S_z, S_z] = 0$. Otherwise, we could not speak of eigenstates as states of a quantum system in which a repeated measurement always provides the same answer. Another example is the space operators, for example $[x, y] = 0$, which means that QM allows for simultaneous determination of a particle's position along the three space-axes, contrary to the case with the spin.

What we have seen so far is one of the basic relations of the so-called '*quantum algebra*'. It differs from classical algebra because in the quantum world, the order of the product terms is no longer arbitrary. In CP, it is normal to think of numbers following commutative laws. For example, no one would doubt that $2 \cdot 3 = 6$ and also that $3 \cdot 2 = 6$. We never give it a thought because we consider it self-evident that the position and momentum of a body or the spin of a planet along all three axes could be measured at the same time. Not so in QM. That is why, in QM, physicists and mathematicians were forced to consider a completely new quantum algebra in a Hilbert space that is very different from the math describing classical Newtonian physics in standard real space. The main difference is that, in

QM, the objects of this algebra describing the dynamical variables are no longer numbers or classical variables but observables – mathematical operators acting on a complex wavefunction or state vector, which are also called '*Q-numbers*', despite not being numbers at all.

This also answers the question of how physicists got to that strange and so-fundamentally-different formal description compared to CP. By taking measurements in the quantum domain, physicists were first led to formulate new laws, which they later encoded into a formal language, not the other way around. It is true that this quantum algebra was developed by great physicists and mathematicians in the 1920s and 1930s, like Schrödinger, Heisenberg, Born, John von Neumann, Dirac, and several others, and has today reached a high level of sophistication. However, its inception came into being not because of the application of common sense and logical rules, but because of rock-solid evidence of what the SG-experiments have shown, as well as of other empiric evidence. Nature imposed it on us. The impossibility of establishing the value of two different quantum dynamical variables, like two spin orientations of an electron, is not a logical consequence of some phenomenon. It is a given fact that has no classical parallel and even no classical logical reason to be. It is as it is. That's also why several textbooks entirely ignore the SG-experiments and pose it instead as basic axioms (an approach the author feels is pedagogically detrimental because it skips the conceptual physics standing behind it and presents QM as nothing more than a mathematical exercise). Therefore, the great service of Stern and Gerlach was not just that of measuring the magnetic momentum of a silver atom, but that of disclosing a weird quantum physical world to us, as well as that of opening the way to a huge mathematical development which turned out to be extremely useful and which is bearing theoretical and practical fruits to this day.

3. Is information fundamental?

The usefulness of SG-devices does not stop there. As an appendix to the previous section, we will focus on an experiment, the so-called '*modified Stern-Gerlach experiment*' (MSG-experiment), which digs deeper into 'quantum ontology'. This experimental setup was not conceived of by Stern and Gerlach but about thirty years later, in 1951, by David Bohm [6]. It is a powerful didactical tool that better highlights some quantum phenomena but is unfortunately explained only rarely in physics undergraduate studies, which go directly into the algebra without establishing a deeper understanding of its meaning and where it comes from.

The basic building block of an MSG can be realized with an apparatus as in Fig. 78. Here, one uses three magnets. The first separates the particle beam, as usual, into its spin components (say, here, the X-axis spin, where the particles with $|+\rangle_x$ spin will follow the upper path, while those with $|-\rangle_x$ spin state will follow the lower one). Then, a second, longer magnet with opposite polarity manages to reverse the deviation paths. Finally, the third magnet stops the trajectories from diverging and merges the two beams into one single output, making them indistinguishable.

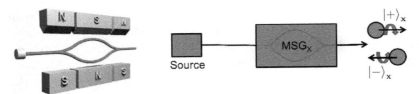

Fig. 78 The modified Stern-Gerlach experiment.

Now, let us repeat the same experiment as in the measurement (D) of the previous section, but with the MSG instead of the SG_x. That is, the experimental situation we have in measurement (E) of *Fig. 79*: The particles in the $|+\rangle_z$ state are selected first and go through the MSG, which deviates it up or down according to its $|+\rangle_x$ or $|-\rangle_x$ spin state. In fact, it physically performs internally the same measure as in the measurement (D) but hides the result to an outside observer by reuniting the two beams at its output.

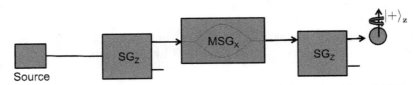

Fig. 79 Measurement (E): Measurement (D) with the SG_x replaced by a MSG_x.

At this point, the question is: What will the last SG_z stage measure, that which repeats the z-spin measure?

We might be tempted to say that there is no reason to believe that a different result than in measurement (D) emerges, as the MSG-device performs an interaction-free measurement along the x-axis and should not change anything along the z-axis.

Moreover, according to our classical logic and understanding of the world, what we know about what happens internally to the MSG is irrelevant. Measurement (E) could be considered equivalent to measurement (D) because we might believe that inserting the MSG simply does nothing, and we would expect, again, a 50-50 chance of getting $|+\rangle_z$ and $|-\rangle_z$.

Nature's answer, however, is that measurement (E) is not equivalent to measurement (D) but to measurement (B). Indeed, the particles all remain unaffected in the $|+\rangle_z$ eigenstate, and remain as they are at the entrance of the whole device! This is despite the fact that inside the MSG_x (that is, hidden from our sight), a sort of 'measurement' along the x-axis is done, just as in measurement (D), with the only difference being that the information it is supposed to hold is erased at the exit of the MSG. Once this cascading of SG and MSG devices is put in place, it does not alter the initial state $|+\rangle_z$ of the particle, apparently in complete contradiction to what we got in measurement (D).

What does all this tell us? There are different ways in which one can describe it using human language.

One way to explain the reality that emerges from this is to state that we perform a spin measurement on a particle along its x-axis and, accordingly, make it travel along the upper or lower path inside the MSG. However, by merging the two particle paths into one, we erase the 'which-way information' the particles followed. It is a sort of 'blind-measurement' in the sense that, physically, the measurement is performed but one does not look at the result and quickly erases the information. Strictly speaking, it shouldn't even be called a 'measurement' because, by definition, this word implies a readout of information (a subtlety that, for a long time, has not been taken seriously and has therefore led to countless misinterpretations and endless confusion to the present day). This was one of the first examples of so-called 'quantum eraser' devices, which we shall see again later in its modern formats. The MSG experiment, therefore, seems to suggest that what really counts is the information we have about the particles. That is also why most physicists tend to believe that the wavefunction represents solely pure information and not a physical entity in Nature.

This sounds quite mind-boggling because, like the wavefunction, the very concept of information is hard to imagine as something 'out there' in the world, as something independent from our abstractions. We think of information as numbers or mental categories, but how can it be something concrete, like a particle's property or the qualities of macroscopic objects, say, of a flower or a stone? Yet, in QP, this seems to be the case. Is it?

Another way to explain the same observed quantum phenomenon is more abstract but it conforms more to the truth. The idea of a quantum particle following distinct paths – here inside an MSG device – is a common fallacy of the human 'sense-mind' which is accustomed to dividing the macroscopic world into distinct objects and which infers its past whereabouts by means of retro-ductive reasoning. There is no which-way information erasure because there is no distinct particle following one or the other distinct paths in the first place, but only something which is in a *'superposition state'* of both possibilities and which is described by a wavefunction, or state vector, that contains the two potentialities at the same time and that evolves without collapse. Only when we measure with a detector (operate with an operator, the observable, on the wavefunction or state vector) by interacting with and eventually absorbing the particle and obtaining a readout (the eigenvalue) does a state projection occur. It would, however, be incorrect to make any statement about the particle's whereabouts before that projective measurement because there was no such thing as '*a particle traveling along the upper or lower path inside the MSG before information erasure*' in the first place. The information is not lost by merging the beams but by detecting or simply blocking them and inducing a projection to only one of the possible eigenstates. We will take this up again later because it is a subtlety that will require more clarification and insight to avoid a very common misunderstanding among even experienced physicists.

Meanwhile, note how from this also emerged the fact that the measurement act in itself did not disturb the systems' state and determine an experiment's outcome. This is also a popular misinterpretation due to Heisenberg's microscope analogy and what we have already dismissed in the section on the uncertainty principle. What ends up leaving the z-spin determined or not is the kind of experimental arrangement that allows for the knowledge about the spin state along the x-axis, or not. Nature forbids us to know the spin of a particle along two axes at the same time. If you know the spin along one direction and then along another one, you will automatically have an indeterminate, an uncertain state of the spin along the former direction, unless you forget about the latter measurement. In that case, everything remains in its nicely defined eigenstate. In other words, QM responds to us depending on the context. One speaks of '*quantum contextuality*', that is, that strange behavior of quantum objects whereby the measurement of a quantum observable depends on the specific experimental setup and the answer depends on what we are allowed to know about it. The result depends not only on the intrinsic physical properties of a quantum object or system but also on the context in which an experiment is done on it. Much more about this will follow.

4. The spinning world of spinors

In introducing some important concepts which will establish the foundation for further discussion and on which we will rely later, it is worthwhile to include a special section that extends the SG experiment to non-perpendicular spin directions.

To sum up, so far, we have considered only the measurement of the particle's spin along the x, y, or z axes, which are mutually perpendicular axes, so called 'orthogonal directions', that is, tilted by a 90° angle to each other. This choice determines the eigenvector basis. We have seen how measuring the spin with a first SG-device along one direction (say, the z-axis) will always give a binary quantized answer – either a spin-up or spin-down momentum, never something in between. Once we measure, say, spin-up angular momentum, we know that this projects the system into the eigenstate $|+\rangle_z$, which means there is probability p=1 (100% certainty) that by taking the same measurement along the same direction, we will find it again to be in the same eigenstate possessing the corresponding same eigenvalue $+\frac{\hbar}{2}$ (in case of fermions, i.e., electrons or protons with half-odd integer spin). Photons have twice as much intrinsic angular momentum, but we also consider them further on; because they have no electrical charge, an SG-magnet can't measure their spin anyway. However, if we subsequently measure the spin observable with a second SG-device along, say the x-axis, we already know that we will get only probability p=0.5 (50% chance) spin-up or spin-down answers. That is, we will find the particles to be, with equal probability, either in eigenstate $|+\rangle_x$ or $|-\rangle_x$, with $+\frac{\hbar}{2}$ or $-\frac{\hbar}{2}$ eigenvalue, respectively.

The natural question at this point is: What happens if, in the second measurement, instead of choosing the x-axis perpendicular to z, we opted for some other axis in between the two, say, at 45°? Half probability and half as much as $\frac{\hbar}{2}$ angular momentum? Or, if we measure the spin along a direction which is only slightly divergent from the z-axis, say, only 1°, will we have still 100% certainty to get $|+\rangle_z$ or a little bit less? If so, how much less?

There is no way to learn this other than experimentally. By smoothly rotating the second SG-magnet, one can measure the spin component, not only along the 90° axes but along any orientation by any arbitrary angle θ, as shown in Fig. 80.

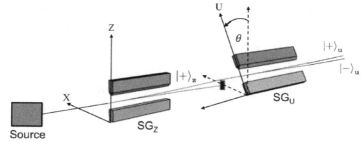

Fig. 80 SG spin-measurement along a different non-orthogonal axis.

Such an experimental setup allows us to measure how the probabilities vary to find a spin-up or spin-down particle (you will never get anything other than such a binary response, whatever angle you choose) by tilting the magnet by an angle θ between the z-axis and a new generic direction. (Let's label it the u-axis.)

The source of particles and the first SG-magnets have only the function to produce and prepare a stream of particles in eigenstate $|+\rangle_z$. Of interest here is what the second SG_u-device will measure along a generic u-tilted axis relative to the z-axis by θ degrees. By doing so, one obtains the probability distribution summarized in Fig. 81.

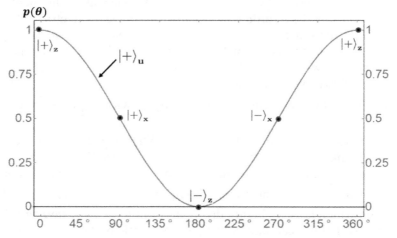

Fig. 81 Probabilities to measure $|+\rangle_u$ or $|-\rangle_u$ spins of a $|+\rangle_z$ prepared particle along the u-axis rotated by an angle θ relative to the z-axis.

Along the abscissa of the graph we can trace the tilting angle θ in degrees of the second SG_u-magnets, while on the ordinate axis we can obtain the probability $p(\theta)$ of getting a spin-up or spin-down measurement result in function of that angle. Let us briefly discuss this in more detail.

Now consider the SG_u oriented as the first one, that is, with angle $\theta = 0$. We are dealing with the same situation as in Fig. 75. Twice measuring the particle in eigenstate $|+\rangle_z$ along the same z-axis doesn't change anything. We will obtain with certainty (p=1) the particle to be in a $|+\rangle_z$ state, whereas with no chance at all (p=0) of getting a $|-\rangle_z$ state particle, even through repeated measurements. Let us now change, only by a little bit, the orientation of the SG_u relative to the z-axis – say, by 1°. Then one would not expect much change. Taking repeated measurements, one will still get in almost all the cases a particle in a spin-up state along the slightly tilted u-axis, that is, the $|+\rangle_u$ eigenstate. However, on a very few occasions, a $|-\rangle_u$ will appear too. After many measurements it will turn out that, on average, only about 7 or 8 particles out of 100,000 will show up with a spin-down state. This corresponds to an almost certain probability for the spin-up of p=0.999924.

Tilting the SG$_u$ further – say, setting $\theta = 45°$ – will definitely change things. About 15% will show up with a spin-down. While, for $\theta = 90°$, which is the measurement of the spin state parallel to the positively directed x-axis, the probability to find one or the other state is equal, that is the 50%, as you should know well by now, and which, of course, is nothing other than what Eq. 21 told us.

Once the SG$_u$-magnets are rotated by 180° to each other, they are anti-parallel to the z-axis and the second SG-magnets look for $|-\rangle_z$ particles. Obviously, though, nothing comes through (p=0) because they were all prepared in the $|+\rangle_z$ eigenstate. Again, for 270° we are in the same situation as that for a 90° x-axis, but with magnets aligned in the opposite direction measuring for $|-\rangle_x$ particles. Finally, a full rotation of 360° brings us back to the original parallel z-axis measurement.

Following are a couple of summarizing considerations which are necessary to avoid later confusion.

First, don't forget that with whatever angle one measures the spin, it will always be a spin-up or spin-down, each of which has an intrinsic angular momentum of $+\frac{\hbar}{2}$ or $-\frac{\hbar}{2}$, but never half or a fraction of it. The rotation of the SG-device changes only the probability of finding the particles in one or the other eigenstate; it will never measure a different angular momentum. Contrary to CM, in QM, from whatever angle you look at it, spin remains a discretized observable.

Secondly, recall that this is only a statistical and probabilistic prediction. This means that only for p=1 or p=0 do you have absolute certainty of observing (or not) a certain spin value, while in all the other cases there could well be statistical fluctuations. For instance, for $\theta = 45°$ you may sometimes have a bit more or less than 15% spin-down particles slipping through.

Where do these random fluctuations come from? In QP, randomness appears to be an intrinsic feature, something with no causes, as if randomicity is without a hidden cause, that is, without '*hidden variables*' (We will return to this important point later, when we will discuss Bell's theorem.)

So, what does Fig. 81 finally show us? It looks wavy, doesn't it? Indeed, we are again dealing with a probability wave! It is like the wavefunction in the double slit experiment. On that occasion, a wavefunction seemed to spread throughout space from the slits and interfere on the detector screen, the squared modulus of which represented a probability distribution. We questioned whether it was a real object or simply a calculation tool. Here, we again have a probability, but we cannot associate it with anything we might imagine as an entity moving in space and time. It is simply the probabilistic description of a particle having a spin-up or spin-down. It turns out that the curve in Fig. 81 is directly proportional to the cosine squared function of half the tilting angle, as:

$$p(\theta) = cos^2 \left(\frac{\theta}{2}\right).$$

In physics, one formalizes this with an elegant formalism that puts all three eigenstates along a preferred z-axis into one single description. For the sake of simplicity, let us restrict ourselves to the z-x plane. Define a spin-up and spin-down along the z-axis, as with a formalism reminiscent of vectors on a cartesian two-dimensional plane, as:

$$|+\rangle_z = \begin{pmatrix} 1 \\ 0 \end{pmatrix} \quad ; \quad |-\rangle_z = \begin{pmatrix} 0 \\ 1 \end{pmatrix} \qquad Eq.\ 24$$

and which are called '*spinors*' to distinguish them from the usual cartesian vectors. In fact, note that these are **not** the basis vectors of the Cartesian 2D plane you learned about in school. (These were $\begin{pmatrix} 1 \\ 0 \end{pmatrix}$ for the basis vector along the x-axis and $\begin{pmatrix} 0 \\ 1 \end{pmatrix}$ for the basis vector along the y-axis.) We are going to see that spinors are quite different objects and require a different treatment – that is, to obey other algebraic rules.

However, note that they represent in any case orthogonal quantum state vectors (recall Eq. 12 and the brief side note on orthogonality in explaining Eq. 18), since:

$$\langle+|-\rangle_z = (1^*, 0^*) \cdot \begin{pmatrix} 0 \\ 1 \end{pmatrix} = 1 \cdot 0 + 0 \cdot 1 = 0 . \quad Eq.\ 25$$

Then, the two Eq. 24, which still represent only two cases (the orientations for $\theta = 0°$ and $\theta = 180°$) can be generalized to any angle in a single state as:

$$|\Psi\rangle = R_y(\theta)\, |+\rangle_z = R_y(\theta) \begin{pmatrix} 1 \\ 0 \end{pmatrix} = \begin{pmatrix} cos\ \theta/2 \\ sin\ \theta/2 \end{pmatrix}, \qquad Eq.\ 26$$

where $R_y(\theta)$ is a rotation matrix which rotates the $\begin{pmatrix} 1 \\ 0 \end{pmatrix}$ spinor about the y-axis of $\theta/2$ degrees. Applying Born's rule – that is, taking the modulus square of the state vector – gives:

$$p(\theta) = |\langle \Psi | \Psi \rangle|^2 = \begin{pmatrix} cos^2\left(\frac{\theta}{2}\right) \\ sin^2\left(\frac{\theta}{2}\right) \end{pmatrix}. \qquad Eq.\ 27$$

This tells us that there is a $cos^2\left(\frac{\theta}{2}\right)$ probability of measuring a spin-up particle along an arbitrary direction rotated relative to the z-axis of θ degrees if it has been prepared in the $|+\rangle_z$ eigenstate, in line with Fig. 81, and equivalently, a $sin^2\left(\frac{\theta}{2}\right)$ probability of measuring a spin-down particle if it has been prepared in the $|-\rangle_z$ eigenstate. The sum of the squared sine and cosine functions always gives, independently from the angle, one $(cos^2\left(\frac{\theta}{2}\right)+sin^2\left(\frac{\theta}{2}\right)=1)$. That is a certainty, as it should be, because it sums over all possible outcomes. That is also why any transformation that transforms a state vector or a wavefunction maintaining the total probability invariant equal to one goes by the name of '*unitary transformation*'. (See also *Eq. 16.*)

Of course, one can extend all this to $|\pm\rangle_y$ eigenstates, but we will not get into this subject. It may only be said that a comprehensive treatment leads to beautiful algebra which paved the way to a phenomenal mathematical development in QT (group theory, representations, and Lie algebra). In QP, spins with its spinors are represented by the so-called '*special unitary group*' in two dimensions, the SU(2) group, and which represents all the transformations that preserve the length of two-dimensional complex vectors. (Don't dare to imagine this; it is a sort of weird three-dimensional rotation inside a four-dimensional space!)

As an exercise, it is instructive to play around with the spinor representation of a spin-½ particle with different angles θ. For example, from *Eq. 26*, for $\theta = 90°$ you get:

$$|\Psi\rangle = |+\rangle_x = \frac{1}{\sqrt{2}}\begin{pmatrix} 1 \\ 1 \end{pmatrix} = \frac{|+\rangle_z + |-\rangle_z}{\sqrt{2}}.$$

This should not come as a surprise. You know that once a particle is in an eigenstate, here $|+\rangle_x$, it must be represented as in Eq. 21, with a superposition state along the other axes.

However, all this implies another weird phenomenon of the quantum world. It is one of the most alien facts to be confronted by our naive human understanding, which is totally incomprehensible for our intuitive daily life experience. Check out what happens when you make a complete rotation, that is, set $\theta = 360°$.

In Eq. 26, for $\theta = 360°$, one does **not** get back $|+\rangle_z$, but obtains $-|+\rangle_z$. Instead, you must rotate it twice to get back to the eigenstate $|+\rangle_z$. Please do not confuse the minus sign in front of the ket, which simply indicates a subtraction operator, with the $+$ or $-$ signs in the state vector which, as usual, stands for the spin-up or spin-down state, respectively. (In fact, note that one could also write it as $e^{i\pi}|+\rangle_z$. See Appendix A I.d, which describes complex numbers, and which denotes a phase shift of the wavefunction by π radians or, equivalently, by 180°; more on this next.)

Then, according to this spinor representation, one must perform a 720° rotation to obtain the same spinor again. Physically, this amounts to saying that electrons, protons, and neutrons, as all '*fermions*', the spin-½ particles, must be rotated by a 720° (or 2π radians) angle to return to their identical initial state. Only with two full rotations will you get back to the original object!

If you are wondering how this could be explained according to an intuitive geometrical understanding, forget about it. It is simply a fact, a physical datum that hardly any normal human being will ever understand. However, mathematically it is a natural consequence of group theory when one works with complex instead of real numbers (mathematicians call it the '*double cover symmetry*' of the SU(2) group) and which again confirms the powerful application of math in physics against all our preconceived ideas of what the world is supposed to look like.

Nevertheless, analogies might help our intuition visualize what is going on here. A typical example of an object with a 720° symmetry is the famous Möbius strip. It is obtained by cutting a band and giving one of the two ends a half twist, then reattaching the two ends. Fig. 82 shows how a Möbius strip changes the orientation after one full rotation. Simply follow the small arrows on the boundary at the tip indicated by the large arrows parallelly displaced on the surface. After a 360° rotation, one does not reach the initial point but travels from the tip to the bottom of the arrows. Another full rotation is needed to return to the original place where the small arrows meet again and regain their original orientation. This is just an analogy that, even though it cannot display the full picture of a four-dimensional rotation, helps one intuitively grasp the spinor rotational property. At any rate, one should refrain from thinking of spinors as being vectors spinning on a unit circle.

One might also ask whether that full rotation that transformed a $|+\rangle_z$ into $-|+\rangle_z$ has any physical significance. After all, what we really measure experimentally is not the state vector but its squared modulus which informs us about the probability of obtaining certain eigenvalues. The sign in front of the Dirac-ket might seem irrelevant because the square of a negative sign disappears anyway.

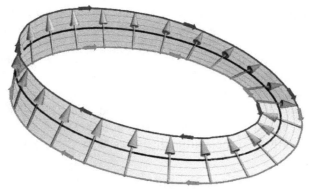

Fig. 82 The Möbius band as an illustration of a spinor rotation.

This belief is, unfortunately, something one finds in textbooks. Not only is this belief wrong, but as we will see later, the seemingly insignificant "-" sign is the manifestation of one of the most fundamental laws of Nature, the Pauli exclusion principle, without which the entire Universe would collapse into an undifferentiated amalgam and perhaps not even exist, at least not as we know it.

In general, a phase shift of a wavefunction $\Psi(x)$ by an angle θ is given by $\Psi'(x) = e^{i\theta} \cdot \Psi(x)$ (see Appendix A I.d). Recall what we have said about the importance of a wave's phase. The destructive or constructive interference between two waves depends on their relative phase. Not only are waves characterized by their amplitude and wavelength (or frequency) but, especially when they are interfering with each other, they must be described using their phase component. The minus sign in front of the full rotated $|+\rangle_z$ eigenstate tells us that it must be thought of as the same state (not to be confused with the $|-\rangle_z$ state, that is, $-|+\rangle_z \neq |-\rangle_z$!) but 180° (or π radians) phase shifted. (In formal terms, one can also write: $-|+\rangle_z = e^{i\pi} \cdot |+\rangle_z$.) Once squared, any phase shift is truncated, and one might believe that it plays no role. However, because phases play an extremely important role in interference phenomena, one might wonder if this strange property of spinors has any physical reality and could lead to physical effects. Or is it just a mathematical abstraction?

The answer is that it is a real physical phenomenon, one observed in neutron interferometry experiments. [7] Neutrons are fermions like electrons but with no electric charge. Therefore, they are easier to handle in experiments involving interference phenomena because they do not interact with the rest of the environment. Moreover, neutrons are about 1838 times more massive than electrons and their de Broglie wavelength is much smaller (when they travel at the same speed), which allows for high-precision measurements. If a neutron is sent through two paths, as was done with photons in the double slit experiment, and one performs a rotation on the neutron traveling along one path (say, with a magnet like in the SG-experiments), the change in the interference pattern will be observed.

Fig. 83 and Fig. 84 illustrate a neutron interferometer which uses three slices of silicon crystals (dashed vertical lines) and which diffracts the neutron's matter waves, as we explained with the Bragg-diffraction experiments. There are no 'paths' in the classical sense, as you should know by now, but only the propagation of spherical probability waves.

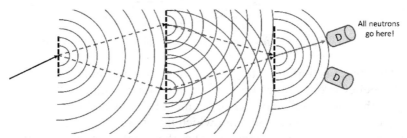

Fig. 83 Neutron interferometry without (or with a 720°) phase shift.

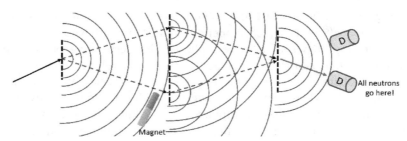

Fig. 84 Neutron interferometry with a 360° phase shift.

However, once a detector clicks, one can (by naive retro-ductive or reverse engineering reasoning) imagine the upper or lower path (oblique dashed lines) as the trajectory the neutron has taken. That is also what you will find frequently in the literature. The neutron's wavefunction incident on the first slice is diffracted by the crystalline lattice, which then is diffracted again at the second crystal slice.

At this second layer, the neutron's matter wave 'sees' the analogue of the Young double slits that photons 'see'. Then, at the third slice, the two matter waves interfere constructively or destructively. Since it is easier to detect the single neutrons with clicking detectors, this third slice works as a beam splitter—that is, it separates the incident neutron beams again in two directions with 50% probability towards the two respective detectors. Assume the two paths having exactly equal lengths. To do this, a precise, angstrom-level (10^{-10}m) alignment of the slices is necessary (technically a quite challenging task, but not impossible).

Fig. 83 shows the case with no phase shift induced on the two paths. An interference pattern forms along the vertical direction parallel to the crystal slices on the plane where the detectors are placed. There, the upper detector, which is placed at an interference fringe, always registers a neutron particle. The other detector, being in a position where no fringe is measurable, never clicks.

In Fig. 84 the same situation is shown but with a magnet inducing a phase shift on one of the possible neutron paths. Magnetic fields always shift wavefunction phases; that is, they cause spinor rotation. This is already strange enough for a neutron, since being an electric neutral particle we don't expect it to have a magnetic moment (recall how magnetic fields are always caused by circulating electric currents). This only signals how the internal structure of a spinning neutron must be composed by electrically charged particles (the quarks), which, however, from the outside cancel each other's field and result to be apparently neutral. At any rate, this is a detail that is not so decisive here. What is essential is that the existence of a magnetic field induces a phase shift on the particles' wavefunction. This phase shift is proportional to a static magnetic field's extension and/or strength through which the particle travels (a SG device also induces a phase shift on the particles it measures). Therefore, by modulating appropriately a magnetic field, one can induce on the neutron beams the desired dephasing on its wavefunction.

It turns out that after a 360° shift, the place on the neutron detection screen where previously (without magnetic phase shift) the bright fringes appeared, is now occupied by dark fringes, and vice-versa. This causes the interchange of the detectors to click. Now it is the second detector that clicks, while the first doesn't reveal any more neutrons. That is, if one shifts by one full turn the phase of the wavefunction, one does *not* obtain the same interference pattern. To obtain the same interference pattern it is necessary to shift the wavefunction's phase by 720°.

This was the experimental proof that was relevant also to the spin phase factor of particles. Later we will find further proof of this in the Aharonov–Bohm effect. Meanwhile we can say, yes, spin-½ particles live in a weird 720° world!

5. The photon's polarization and spin

The cosine squared spin orientation probability distribution emerging from the SG-experiments did not really come as a surprise to physicists. Something quite similar was well known much earlier than the advent of QP by observing the behavior of polarized light.

We already hinted at the property of polarization that an EM wave can have. In Fig. 4 a vertically polarized EM wave was depicted, which means that the vector of the propagating electric field oscillates along a vertical y-axis, perpendicular to its direction of z-propagation. Usually light from natural sources is not polarized; there exists no preferred plane along which the electric field of a beam of light is polarized, while it is allowed to oscillate along all possible directions. An unpolarized source of light emits its EM waves randomly, as shown for instance in Fig. 85 left.

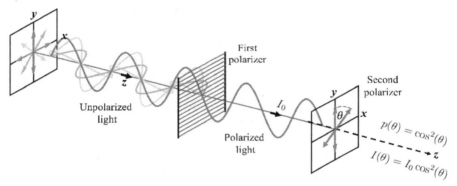

Fig. 85 Linear polarization of light and polarizer filters.

However, it is possible to filter out polarized light from an unpolarized source. Polarizers are optical filters that allow only a specific polarization to pass through and absorb or reflect any other (not to be confused with wavelength filters that select specific wavelengths or wavelength intervals). They could be made of special crystals, glass plates, thin films, metallic wire grids, and so forth, which can all perform a polarizing function. Polarizing filters are used in camera lenses in photography or in sunglasses to avoid unwanted reflections, darken skies, suppress glare from the surface of a body of water, and so forth. A relatively simple type of polarizer is the wire-grid polarizer made of many fine parallel metallic wires on a plane, which filters out a linearly polarized EM wave with the electric field perpendicular to the wires, as shown in Fig. 85.

Once polarized light of intensity I_0 is obtained with this first filter, a second polarizer can be added to dim the polarized beam. The first polarization of EM waves, or photons, is analogous to the selection performed by the first SG-device on the electrons, along a specific orientation. However here, instead of selecting out the electrons with the spin-up eigenstate, it is the photons with a specific polarization (not their spin) which are filtered out along one preferred direction. Analogously, the second polarizer acts in a way similar to that which we saw with the second SG-device on electrons. If it is oriented along the same direction of the first polarizer it simply lets the same photon in its polarization eigenstate pass through with 100% probability. The question then is whether and how the 'dimming factor' dependent on the rotation of the second light polarizer for photons is comparable to the squared cosine probability law we discussed earlier for the electron's spin.

Again, the answer cannot be derived from logical reasoning but must come from empirical data. By measuring how the intensity I of the outcoming beam depends on the relative orientation by an angle θ of the two polarizers, one obtains the graph of Fig. 86.

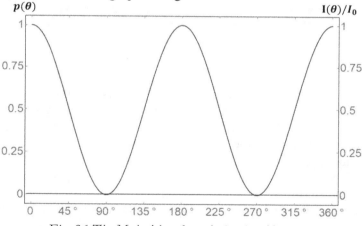

Fig. 86 The Malus' law for polarization filters.

Also, this time the intensity varies according to a squared cosine law, but—and this is of paramount importance to keep in mind—with a θ instead of a $\frac{\theta}{2}$ dependence! For instance, it takes a 90° rotation to stop the EM beam, instead of the 180° necessary for the electrons (compare Fig. 81 and Fig. 86). This is summarized by the Malus' law:

$$I(\theta) = I_0 \cos^2(\theta)$$

It can be seen as a natural consequence of the fact that the polarizer simply selects out only the vertical component of the electric field vector tilted by θ degrees relative to the vertical axis. By simple trigonometric considerations, the cosine of the original wave amplitude remains (see the diagram of Fig. 179 in the appendix for help), and to obtain the intensity one has to square it. So, there is nothing particularly 'quantum' in this phenomenon.

It is two-centuries-old knowledge that light can be polarized; the French physicist E. L. Malus (1775-1812) discovered its law in 1809. But, considering what has been discovered in the time since, especially comparing photons with the fermions behavior, we can recognize interesting facts about both. Let us translate what we have acquired so far into a corpuscular picture.

Fig. 87 Étienne-Louis Malus (1775-1812)

Since the probability of a photon going through a polarizer is nothing else than the proportionality factor of the number of transmitted photons I versus the incident photons I_0 ($p = I/I_0$), Malus' law amounts also to saying that the probability of a single photon passing through the second polarizer is:

$$p(\theta) = cos^2(\theta). \quad Eq.\ 28$$

Nowadays, this can be checked directly. It is possible to dim the light source to such a degree that only one photon at a time goes through the polarizers (recall how we have already seen this for the double slit experiment), and count the single photon with a 'clicking detector' on the other side one at a time. By doing so, one discovers that, in fact, the probability which the detector clicks follows a cosine squared law. With this we reestablish the wave-particle duality again. One can take the classical point of view of the classical physicists like Malus did, who assumed light to be a wave. The polarizer does not block the wave entirely (except of course for a 90° orientation) but simply dampens it. Whereas from the corpuscular perspective, there is no damping—only cutting the number of photons allowed to get through the polarizer without reducing the single photon's energy. The photon passes or does not pass (i.e., is absorbed or eventually reflected), but you will never register half a photon. For 89.9° there are, on average, still three photons out of a million which will pass the polarizer but, once they will have managed to go through, will be exactly like the incoming ones.

Along our journey we will later discover how the Malus law will lead to one of the most unexpected consequences for our worldview. It is the death sentence for any local and realistic hidden variable theory. But until the

1960s, this was something far from obvious. Other principles, facts, and foundational issues had to be clarified first.

An important question at this point is: Why do we measure for photons a θ instead of a $\frac{\theta}{2}$ squared cosine dependence, as we saw for electrons?

Recall that for the intrinsic momentum of an electron one gets through any measurement of the spin component, that is, the eigenvalue associated to the eigenvector state, always $S_{e^-} = \pm\frac{\hbar}{2} = \pm\frac{h}{4\pi}$ where 4π is nothing other than a 720° angle in radians. Planck's constant h is a natural constant and is derived experimentally with the photoelectric effect, and the spin S_{e^-} is also not something that can be derived but must be measured experimentally, for example with a SG-device. The fact that just h and S have an inverse 4π proportionality, which is related to the 720° rotation reflected in the squared cosine of $\theta/2$ instead of θ, is not a mere convention but reflects some deep principle that is imposed by Nature. Therefore, since for photons we have the same cosine squared intensity, but for a θ dependency, that is a normal 360° full rotation, it is a plausible assumption that the photons' intrinsic angular momentum amounts to $S_\gamma = \pm\hbar = \pm\frac{h}{2\pi}$. Or, in other words, photons do us the favor of living in a less weird world; they perceive rotations as we humans do. One might also say that polarizing sunglasses provide proof that photons are spin 1 particles and live in a 360° world.

More generally, one can state that the spin value of a particle reflects the kind of rotational realm it lives in. We can extend this reasoning by predicting that spin 2, spin 3, etc. particles need only 180°, 120°, etc. rotations to bring them back into the same quantum state, while spin $\frac{3}{2}$, spin $\frac{5}{2}$, etc. particles need 240°, 144°, etc. rotations to look the same. Finally, spin 0 particles never change for any angle one chooses to look upon them, since they have no momentum vector in the first place.

However, Nature seems to restrain itself, since all known particles are spin $\frac{1}{2}$ (fermions—that is, leptons and quarks) or spin 1 (most bosons, for instance the photon) and spin 0 (the Higgs boson). The gravitons, that is, the particles supposed to mediate the gravitational force, are predicted to have spin 2, while other speculative particles (like those predicted by supersymmetry theories) are believed to possess other spin values eventually as well, but these have not been detected so far, despite intense research. The world we perceive with our human senses is therefore dominated almost exclusively by spin $\frac{1}{2}$, spin 1, and spin 2 particles (the Higgs boson has lifetimes of the order of 10^{-22} seconds and is way beyond human perception).

But what about the photon's spin? Another main difference between the spin of massive particles and that of massless particles such as a photon is that the latter has not only two possible spin states but also only two possible spin orientations: along the direction of propagation or opposite to it. A photon has a *'helicity'*, which tells us whether the spin is aligned or anti-aligned with the momentum (see Fig. 88).

In general, a particle is said to be right-handed (i.e., have positive helicity) if the spin and momentum are parallel, while it is said to be left-handed (i.e., have negative helicity) if the spin and momentum are antiparallel. So, each individual photon will have its spin either along the direction of motion or against it. Other spin directions are not allowed.

Fig. 88 Positive helicity (right-handed) and negative helicity (left-handed) particles.

The conceptual connection between waves and photons with their positive or negative helicity is realized by the electric field rotation of circular polarized EM waves. Circular polarization means that the direction of the electric field vector rotates with time at a steady rate in a plane perpendicular to the direction of the wave propagation. For right circular polarization, the electric field vector rotates in a right-hand sense with respect to the direction of propagation (see Fig. 89 left), while for left circular polarization, the electric field vector rotates in a left-hand sense (see Fig. 89 right).

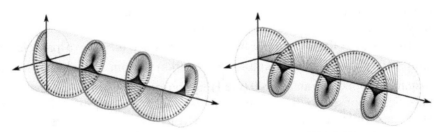

Fig. 89 Right and left circular polarization of an EM field.

The photons of counter-clockwise right-polarized light carry an angular momentum of $S_\gamma = +\hbar$ along the direction of flight. When light is clockwise left-polarized, then photons carry an angular momentum of $S_\gamma = -\hbar$ along the flight direction.

Note how in the case of linearly polarized light, photons can be considered to be in a state of superposition of the right and left circular polarization. What that precisely means will be clarified in the next section, which is dedicated expressly to the principle of quantum superposition.

IV.   Quantum ubiquity and randomness

1. The quantum superposition principle: being in two states at the same time

Let us focus now on another very important concept of QP, which is the so-called '*quantum superposition principle*'. It is a very interesting but also weird aspect of quantum reality which has been discussed by scientists and philosophers throughout the twentieth century and still remains a deeply counterintuitive property of particles—and not only particles.

So, let us see what it is all about by looking back briefly again to the good old double slit experiment. You recall how on that occasion we conceived of an experiment where we could use only one particle at a time, say a single electron at a time, which we sent through the double slits towards the screen. Then, by repeating these single particle shots, after a sufficient number of electrons were collected on the detection screen, we obtained the typical diffraction and interference pattern. We concluded that the interference phenomenon, which emerges even if only one particle at a time is involved, cannot be explained if we conceive of the electron as a single point-like particle that goes through one or the other slit, but we must think of it as an extended wavefront that goes through both slits or a particle going through one *and* the other slit at the same time. It is hard enough to imagine a single particle diffracting and interfering with itself, but if we stick with this picture, we must give up the logical OR reasoning. We must think of the particle going through slit 1 *and* slit 2, something that in CP, as well as classical logic, would be mutually exclusive. As weird as this may appear, experimental facts show that our intuitive understanding, which presupposes an interpretation based on a logical disjunctive OR, would be in contradiction with the observed interference phenomena.

Since this state of affairs sounds counterintuitive and the model of a single particle becoming ubiquitous and going through two or eventually more slits is a quite doubtful one, physicists refrain from trying to imagine any model and ontology at all. One then simply says that the state of the particle, as long there is no measurement, is 'undetermined', 'uncertain', or simply 'undefined'.

We encountered a similar situation with Heisenberg's uncertainty principle, which we visualized with the help of a single slit or the pinhole experiment, and which is summarized in Fig. 90.

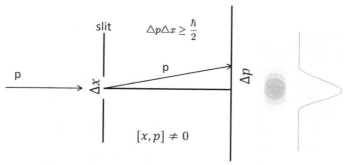

Fig. 90 Heisenberg's uncertainty principle as a
superposition of states of the particle's location.

What we observe on the screen is the distribution of points, of bright tiny spots on the screen, which statistically can be represented by a bell-shaped curve centered on the slit aperture. When we send many particles one after another, after some time a bright fringe builds up, which we can describe by this bell-shaped distribution function (yet, eventually displaying tiny secondary interference minima and maxima which remind us that the wave nature is still present, even though most of the time it is negligible). This tells us where the single particles hit the screen, accumulating one after the other in a region around the slit's center and which describes our uncertainty on the whereabouts of the particle that formally is given by the Heisenberg's uncertainty relation of Eq. 8 or the corresponding commutation relation of Eq. 23.

This means that, before the measurement, we can interpret the particle as being in all places at once when it traverses region Δx of the slit size, and, once it is in front of the screen, which is shortly before the wavefunction 'collapses' to a point, it is everywhere simultaneously along the space interval described by the distribution function. Considering that the position in space of a particle is also a property that describes its quantum state (recall that in QM the position is in fact obtained via the position observable, which is the x operator), this suggests a more general interpretation: Namely, that before this instant of interaction of the particle with the detector, we must consider it in a 'superposition' of states. Before the interaction the particle's location in space is not just unknown, it is in a superposition of several possible locations.

That is also why we cannot interpret Heisenberg's uncertainty as something due to our ignorance about the position of the particle in terms of a logical OR reasoning ('the particle is in one or the other state'). It must be interpreted as logical AND as a co-existence of states, that is, with all possible states superimposed ('the particle is simultaneously in many possible states'). QM should be taken seriously. In fact, it challenges us to believe that a physical object can be in two, or even in many, possible places at the same time—or in no physical state at all!

This strange 'ubiquity' of particles conveys a deeper meaning to the commutation relations. The fact that observables x and p do not commute means not only that once we try to determine x with high precision the values of p get 'smeared', but that this implies also the existence of state superposition: If two dynamical variables are represented by two non-commuting operators, once a measurement of the first is done, a superposition of states corresponding to the other dynamical variable is automatically induced. This suggests also how this weird superposition effect must not be exclusive to space but is a much more general principle of QM, since there are also other observables that do not commute.

For instance, as you will remember, the spin does not commute. Indeed, we have already seen something similar in the SG-experiments. As the commutation relations of the position and momentum operators are a direct consequence of Heisenberg's uncertainty principle, which can be illustrated with the pinhole experiment, so the commutation relations of the spin operators are a direct consequence of the SG-experiments. The only difference is that the observables x and p are allowed to acquire any value, as they are continuous quantities, whereas in QM the spin-state of a particle can acquire only two values, which we labeled as the '+' and '-' or the 'up' and 'down' intrinsic angular momentum. And, once we measure the particle spin along an axis, say the z-axis, automatically the spin states along the other axes are not just uncertain but in superposition. In QP, the concept of the state of 'uncertainty' must be interpreted as superposition of states, not as an ignorance of which state is realized.

Let us visualize and define this a bit more rigorously with the help of Fig. 91 (inspired by the example of measurement (D) in *Fig. 77*).

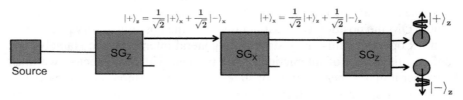

Fig. 91 State vector representation along different bases (example of measurement (D) - Fig. 77).

As usual a first SG$_z$-device prepares a particle in a $|+\rangle_z$ eigenstate. Here, to 'prepare' means that those particles showing up with the spin-up state along z were first selected from a source by performing a measurement that projects them into the $|+\rangle_z$ eigenstate, that is, the S_z observable comes into the play first. Since the S_z operator does not commute with the S_x and S_y observables (see Eq. 22), this measurement automatically renders the particle's spin state along the x- and y-axis as uncertain. However, 'uncertain' does not only mean that, once measured, there is a 50% chance of finding the particle in a $|+\rangle_x$ or $|-\rangle_x$ eigenstate, but that before the next measurement it is in both states: The $|+\rangle_x$ *and* $|-\rangle_x$ states coexist. Before such a measurement takes place, therefore, we can represent the state (not the eigenstate) of the spin-up particle along the z-axis, $|+\rangle_z$, in superposition of both spin-up and spin-down along the x-axis basis as:

$$|+\rangle_z = \frac{1}{\sqrt{2}}|+\rangle_x + \frac{1}{\sqrt{2}}|-\rangle_x \quad Eq.\ 29$$

This is not just a mathematical description which furnishes the probabilities of a result. That little tiny '+' in Eq. 29 (or Eq. 21) represents a superposition of spin eigenstates. Whereas, as usual, the square rooted coefficients tell us something about the probability of getting one or the other result, should an x-directed measurement take place. The two particles' intrinsic momenta along the x-axis (in this example similar considerations would hold for the y-axis, as well) are both present, as if the particle is spinning clockwise AND anti-clockwise at the same time. Not only do quantum particles multilocate, they even 'multispin'!

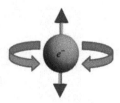

Fig. 92 What is a particle in spin-superposition?

Then, once the second measurement has taken place, which projects the particle's spin along the x-axis with the SG$_x$-device, the x-spin has been measured and the spin-up eigenstate $|+\rangle_x$ selected (the $|-\rangle_x$ eigenstates are blocked). The superposition along the x-axes has gone; vector state reduction has occurred. But since the spin operator S_x does not commute with S_z, the particles' state along the z-axes is now no longer in eigenstate $|+\rangle_z$. The spin-up particle's eigenstate along the x-axis basis, $|+\rangle_x$, can be represented as being in superposition of both spin-up and spin-down along the z-axis basis as:

$$|+\rangle_x = \frac{1}{\sqrt{2}}|+\rangle_z + \frac{1}{\sqrt{2}}|-\rangle_z \qquad Eq.\ 30$$

Finally, the third stage device in z selects out either the $|+\rangle_z$ or the $|-\rangle_z$ state, as amply discussed in the section on the SG-devices.

Therefore, the same quantum state $|+\rangle_z$ represented in the z-basis can be represented in the x-axis eigenvector basis as well. In a similar fashion quantum algebra allows for the representation of a quantum state in any desired eigenvector basis (such as the y-axis or any basis in between). It is like 'seeing' the very same quantum state from different geometric perspectives by a change of coordinates, for example with a rotation or translation, but leaving it physically unaltered.

So far, we have considered particles with a mass, but another example where quantum superposition takes place is the superposition of photon states. Photons are massless particles and, in vacuum, always travel with the speed of light, which is about $3 \cdot 10^8$ m/s, whatever the speed and dynamic state of the observer or the measurement apparatus (this is one of the main principles of SR). You will never 'see' a photon at rest as you do in the case of particles possessing a mass. We already saw that the positive or negative spin of a photon is oriented along or contrary to the direction of propagation and is determined by the right- or left-handed rotation of circular polarized EM waves. What polarizers select out, however, are not spins, but the photon's electric field orientation. This doesn't mean that photons can't be in superposition state. As particles can be in two places at once, or have two spins at once, they could have also two polarizations at once.

We can apply a similar reasoning to photons as we have done to electrons, provided we remember that the former are particles with spin 1 directed only along or contrary to the momentum direction (contrary to fermions, which are particles with only two possible spin-½ outcomes that, however, can point towards any direction), that the polarization vector is restricted to an x-y plane perpendicular to the direction of propagation (the z-direction) and that photons are not described by spinors but by the EM field vectors which return onto themselves after 360° rotation (that is, all angles here are half as much as those of fermions).

Then let us agree on how to use certain symbolic conventions. Since in the literature this can be quite different, let us be flexible and use different symbolic conventions according to the context. We will use the following state vectors notation.

For the up-down vertical polarization state vectors: $|\uparrow\rangle = |V\rangle$ and $|\downarrow\rangle$.

For the right-left horizontal polarization: $|\rightarrow\rangle = |H\rangle$ and $|\leftarrow\rangle$.

For the diagonal 45°, 135°, 225° and 315° polarizations: $|\nearrow\rangle$, $|\searrow\rangle$ $|\swarrow\rangle$ and $|\nwarrow\rangle$ or $|45°\rangle$, $|135°\rangle$ $|225°\rangle$ and $|315°\rangle$, respectively.

For a generic polarization of θ-degrees: $|\theta°\rangle$.

Here let us use first the graphically more intuitive arrow-convention.

Then we can reason by analogy and say that, as an electron's spin selection along the z-direction changes its x-spin components into a x-spin-superposition state (that is, Eq. 29), so a photon's polarization (again, not to confuse with the photon's spin) along the vertical y-direction selected from an unpolarized light source superimposes its -45° and +45° polarization components into a diagonal polarization-superposition state as (see Fig. 93 left):

$$|\uparrow\rangle = \frac{1}{\sqrt{2}}\,|\nwarrow\rangle + \frac{1}{\sqrt{2}}\,|\nearrow\rangle. \qquad Eq.\ 31$$

Notice how we used the words 'selection' and 'change' which do not necessarily involve the state reduction of a measurement.

Similarly, one can represent the diagonal polarization as a superposition of the vertical and horizontal one as (see Fig. 93 right):

$$|\nearrow\rangle = \frac{1}{\sqrt{2}}\,|\uparrow\rangle + \frac{1}{\sqrt{2}}\,|\rightarrow\rangle.$$

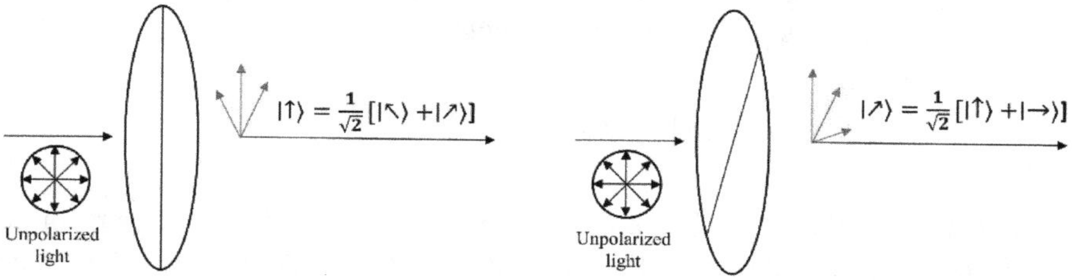

$$|\uparrow\rangle = \frac{1}{\sqrt{2}}[|\nwarrow\rangle + |\nearrow\rangle] \qquad\qquad |\nearrow\rangle = \frac{1}{\sqrt{2}}[|\uparrow\rangle + |\rightarrow\rangle]$$

Fig. 93 Superposition of photon polarization states.

Or, as another example, the horizontal polarization vector can be represented as the sum of the 45° and 135° diagonal vectors as (not shown):

$$|\rightarrow\rangle = \frac{1}{\sqrt{2}}\,|\nearrow\rangle + \frac{1}{\sqrt{2}}\,|\searrow\rangle. \qquad Eq.\ 32$$

Of course, many other combinations are possible. After all, that is nothing else than the vectors summation according to the parallelogram law (see Appendix A Ib) and doesn't look particularly surprising. Polarization vectors, as any other physical vector quantities, such as speed, accelerations, forces, etc. are treated with a vector formalism. In QP, however, this acquires a deeper significance. A quantum state superposition is not just the addition of two states resulting into another state but the realization of both states simultaneously.

This suggests introducing for photon polarization a more general vector basis description as we have already done with fermion particles (see Eq. 26), as follows:

$$|V\rangle = \begin{pmatrix} 1 \\ 0 \end{pmatrix} \; ; \; |H\rangle = \begin{pmatrix} 0 \\ 1 \end{pmatrix}; \; |\theta\rangle = \begin{pmatrix} \cos\theta \\ \sin\theta \end{pmatrix}$$

Where the first basis vector is the vertical polarization, the second basis vector the horizontal polarization, and the third a generalized version polarization state vector in dependence of the angle θ, which, given as a superposition of the two bases vectors, can be formulated:

$$|\theta\rangle = \cos\theta \, |V\rangle + \sin\theta \, |H\rangle$$

Note that this time the cosine and sine functions have θ instead of $\frac{\theta}{2}$ dependence, which means photons return to themselves after rotations of 360°, and the world behaves nicely again!

Finally, applying Born's rule as we have done for spinors (see Eq. 27), taking the modulus square of the state vector gives:

$$p(\theta) = |\langle\theta|\theta\rangle|^2 = \begin{pmatrix} cos^2\theta \\ sin^2\theta \end{pmatrix}$$

This tells us that there is a $cos^2\theta$ probability that a photon slips through a polarizer rotated relative to the vertical y-axis by θ degrees if it has been prepared in the $|V\rangle$ eigenstate, and equivalently, a $sin^2\theta$ probability to measure a photon if it has been prepared in the $|H\rangle$ eigenstate.

To sum up and conclude this tour on the quantum superposition principle, what we should keep in mind is the bottom line of all this: At least for non-commuting observables, it makes no sense to say that a particle is in one or another state until a measurement has projected it into an eigenstate. We can only say something about the probability to measure one or another outcome—and not because of ignorance, but because the system is intrinsically in a superposition of states until you measure it. That is, only once the measurement occurs, in the case of Heisenberg's uncertainty, does the wavefunction instantly 'collapse' or, in the case of the spin measurement, the state vector reduction (or projection) on its eigenbasis occurs. This collapse or state reduction happens according to a totally random law (given by the squared modulus of the wavefunction or that of the coefficients of the state vector). And, in the formal representation of a superposition, like that of Eq. 29, Eq. 30, or Eq. 31, we must read the plus sign as a logical conjunction of two co-existing states.

The single and double slit experiments and the SG-experiments were only a couple of examples which tried to shed some light on a general fact in QP. Many other experimental approaches would have been possible, but they would all lead to the same conclusion: that it is perfectly admissible to have a system taking two or more possible states at once.

All that might look impossible or unappealing to some, but what is definite is that once you use the quantum superposition in the mathematical formalism of QM, then everything works fine. Only if the superposition principle is taken seriously will the match between the theory and the experimental check with reality turn out to be correct. Later on we shall encounter other experiments which will underline this further.

2. The time-energy uncertainty relation

Apart from the position-momentum Heisenberg uncertainty relation, there exists also a time-energy uncertainty relation. With Heisenberg's uncertainty relation $\Delta x \, \Delta p \geq \frac{\hbar}{2}$ there is likewise a so-called 'time-energy relation':

$$\Delta E \cdot \Delta t \geq \frac{\hbar}{2} \qquad Eq.\ 33$$

with Δt being the time interval it takes to perform a measurement or the time interval a physical phenomenon takes to occur (for example, an atomic energy transition) and ΔE being the uncertainty over the energy of the measured system. The time-energy uncertainty relation tells us something about the shortest average time Δt we need in order to notice a change in energy of a quantum system by an amount equal to its standard deviation ΔE. Stated more simply, the faster a measurement is performed, the greater the uncertainty on the energy readout.

The time-energy uncertainty should not come as a surprise. Since there is always an uncertainty over the momentum of a particle, this also implies an uncertainty over its energy, because its momentum and kinetic energy are two directly related physical quantities. And since space, time, and energy are the three most important ingredients with which the physical Universe is made of, it is clear that these two simple little inequalities are expressing something very fundamental about our physical existence. Just as in the case of Heisenberg's uncertainty, the time-energy uncertainty arises from the wave properties inherent in the quantum mechanical description of Nature as well.

In modern textbooks on QM, this uncertainty relation as well as Heisenberg's uncertainty relation are usually both derived from operator algebra and statistical considerations. We don't give these rigorous mathematical proofs here, but let us attempt an intuitive approach that can illustrate where this relation comes from. Let us limit ourselves to the case of a photon of energy E. We know from Eq. 7 that it carries a momentum $p = \frac{E}{c}$. Its uncertainty over the momentum Δp due to the uncertainty over the energy ΔE can therefore be written straightforwardly as $\Delta p = \frac{\Delta E}{c}$. Insert this into Heisenberg's uncertainty relation and you will get the time-energy relation if one imposes $\Delta t = \Delta x/c$. This latter equation expresses the time that light needs to travel through the space interval Δx. Then, the quantum uncertainty of a photon's energy during a time span Δt is $\Delta E \geq \frac{\hbar}{2\,\Delta t}$. Or, equivalently, we must wait a time $\Delta t \geq \frac{\hbar}{2\,\Delta E}$ to notice a change of the photon's energy by an amount ΔE.

In words, that means that the uncertainty over the energy of a quantum state is inversely proportional to the time scale during which it occurs or during which we observe it. The shorter a measurement is (or, the shorter the time during which a quantum event occurs), the larger becomes the possible deviation from an average energy value we will measure and the more uncertain and fuzzy the energy of the events we will observe becomes. So, if we desire to obtain an infinitely precise measurement of the energy of, say a particle, we would have to wait an infinitely long time to obtain the result of the measurement. And, it can't be emphasized enough that this is not due to the inaccuracy of our measurements but rather is an intrinsic property of Nature. From the practical point of view, even in principle and with the most ideal and infinitely precise measurement device, we will always be bounded by this natural and intrinsic uncertainty of the single outcome at every measurement. Or, to see it from another perspective, one needs much more time to prepare a system in a sharply defined energy state than one with a less determined one.

However, for the single photon that doesn't mean that one never will measure its energy with less uncertainty. It is all about a statistical fluctuation, an average uncertainty one should expect, independent of the perfection and precision level of the measurement apparatus.

To clarify this important conceptual aspect let us take a step further and be more precise. First of all, the symbol Δ refers to the statistical standard deviation. We already saw how this is the statistical fluctuation of a set of measured values around some average value, with the latter being the expectation value we discussed in Chapter II.9. In the case of interest, here it tells us how large the energy standard deviation ΔE is, as a departure from average energy $\overline{E}$ obtained ideally after an infinite number of measurements.

Once you have measured a sufficient number of events you should obtain the typical bell-shaped exponential probability density function, that is, the 'normal distribution' (or Gaussian probability density function, as of Fig. 60), which we apply again to illustrate energy fluctuations around a mean energy in Fig. 94.

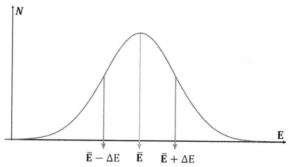

Fig. 94 Standard deviation around the mean energy.

Here, we have on the abscissa the energy and on the ordinate axis the number of obtained measurements corresponding to some energy E_i. As a rule of thumb, the probability to find an energy measurement inside the standard fluctuation ΔE around a mean value $\overline{E}$ is about 68%. This means that we can observe events with more uncertainty than ΔE, but with less probability.

The similarity between Heisenberg's uncertainty principle and the time-energy uncertainty should not mislead, however. As we have seen, in QM operators can represent the position and momentum, as their complementarity can be expressed by non-commuting operators. We have seen that the energy of a system can also be represented by an observable, an operator which acts on the wavefunction, and which expresses the Schrödinger equation, the Hamiltonian operator H. However, there is no time operator in QM, and therefore there is also no commutation relation between time and energy. In QM, time appears only as a parameter in the equations.

It turns out that time is not an observable because it is not something we can measure. This might sound strange at first. After all, in our vernacular we speak of clocks as 'measuring time'. But if you think about it, you realize that clocks measure a change in a physical system and indicate it by an arrow or digital display. Time is too subjective and indefinable an entity for anyone to ever really measure, in the sense that you can measure the position of a particle or the length of a table. This is because time is not something inherently possessed by an object, as are its length, mass, position, energy, or charge. It is something we use a posteriori as a measure of change of these inherent properties, but it is not a property of objects themselves. The quantum world reminds us of this: There is no such thing as a time operator. Time is a parameter, just a number—that is, a scalar quantity.

One can compare this with the space-time notion of relativity. Here, time and space are on equal footing. Physicists even use time as a fourth dimension together with the three spatial dimensions, whereas in QM, time appears as a very different, separate entity from space.

Heisenberg's (space-momentum) uncertainty relation $\qquad \Delta x \cdot \Delta p \geq \dfrac{\hbar}{2} \qquad [x, p] = i\hbar$

$\Updownarrow$

Time-energy uncertainty relation $\qquad \Delta E \cdot \Delta t \geq \dfrac{\hbar}{2} \qquad [H, t] \neq i\hbar$

*Fig. 95 In QP time is **not** an operator!*

There is something mysterious in this discrepancy that is debated even nowadays among philosophers and physicists. It is also one of the stumbling blocks for modern physicists who are trying to unify QT with general relativity (GR) into a unique unified theory. One reason for why this turns out to be so difficult is that no one knows how to build a theory that brings under the same umbrella such diverse conceptions and expressions of space and time as we find in GR and QM.

So, let us consider an example where the quantum time-energy uncertainty plays an important role. An instance where this occurs is that of the atomic energy transitions. In Fig. 96 we can see the example of the hydrogen atom photon emission. Here the transitions occur very quickly, over a time of $\Delta t < 10^{-10}$s (less than a tenth of a billionth of a second). This means that, according to the time-energy uncertainty relation, every time an atom emits a photon, it will not always possess the same energy as the previous one.

Even if for each single individual photon one measures a definite energy, the subsequent measurements of several other photons will give a slightly different energy E for each, with a random distribution centered on the mean, that is, with $\overline{E} \pm \Delta E$, $\overline{E}$ the average atom's energy and ΔE its standard deviation given by $\Delta E \geq \dfrac{\hbar}{2 \Delta t}$. You might even detect photons with even greater energy deviations than ΔE, but the probability that this will occur decreases quickly (i.e., exponentially) according to a normal distribution Gaussian law.

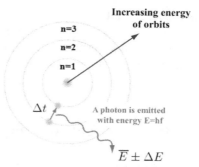

Fig. 96 The time-energy uncertainty in atomic transitions.

This implies that the lines of atomic spectra are always a bit 'smeared' over an energy interval. In conventional spectrographs the lines are just the projection on a screen of a slit through which the light source shines (for instance, see in Fig. 13 the hydrogen's and mercury's spectral emission lines). There will be two causes of the lines' thicknesses. The first cause, you should already know, is the diffraction effects of the slit (or equivalently, due to Heisenberg's uncertainty). The second cause, which adds up to the previous one, is the atomic levels energy uncertainty. We leave it to you, as homework, to show that for a pure yellow light source (say a laser) of average wavelength of 600 nm ('nanometer;' 1 nm= 10^{-9} m), with $\Delta t = 10^{-10}$s transition time, the time-energy relation leads to an uncertainty of about 10^{-3}nm. This is only a tiny departure from the mean wavelength of the source and is negligible for most applications. But in some applications which require high precision, these effects must be taken into account.

To sum up, time and energy are intimately related in QM in such a way that it makes no sense to speak about precise and definite energy values. Everything depends on the timescale we are considering. If a quantum event occurs on very short time scales (e.g., nuclear decays occur at the order of 10^{-25} seconds!) its energy will inevitably be undetermined, fuzzy, and smeared out over an interval of energy values. Again, we see that it makes no sense to speak about objects as having an objective definite precise physical quantitative property, such as here the energy at a specific instant in time, no more and no less than it makes sense in QM to speak of things as possessing both a precise position and a momentum. 'Objectivity' has become a very questionable thing!

The question that should arise naturally at this point is: Where do these 'fluctuations' come from? Are there really things in the quantum world 'fluctuating' out there? Where does that extra (or missing) energy content ΔE come into play (or go to) if the universal law of energy conservation is supposed to hold? Is it energy that pops into existence from out of nothing? We will try to answer these important questions in the coming section on zero-point vacuum energy, virtual particles, and the Casimir effect. Before that, however, let us take a look at another strange effect related to the time-energy uncertainty: the tunnel effect.

3. The quantum tunnel effect: the impossible jumps

Another interesting effect that you will find in the realm of the quantum world is the so-called *'quantum tunneling effect'*. It is a strange phenomenon that emerges quite naturally once the time-energy relation, the Heisenberg's uncertainty relation, and the wave-particle principle expressed by the de Broglie relation are taken together. We saw that the time-energy uncertainty tells us that the energy of a particle is intrinsically undetermined and that this lack of definitiveness depends on the time interval during which the process or the observation takes place. This is very different than in Newtonian physics, where everything is considered to have a precise and definite energy (be it kinetic or potential, chemical, mechanical, electric, or whatever kind of energy), and where the only incertitude arises due to our ignorance and the not ideal measurement devices, which are always affected by some sort of error and imperfections. This implies that we have to expect very different behaviors between classical mechanical and quantum objects.

For instance, according to CM we can throw an object over a wall only if it has a sufficient kinetic energy that can overcome the wall, which means, more precisely, that there is enough kinetic energy that can be transformed into potential energy to reach the height of the top of the wall. If it does not have that or more speed in the form of kinetic energy to overcome this potential barrier it will hit the wall and eventually bounce back by a more or less elastic or inelastic reflection. However, in QM, things are a bit different. Any object's position and energy is, due to Heisenberg's time-energy uncertainty relation, always a bit uncertain, intrinsically fuzzy, and smeared out over an average value (in Fig. 97, for simplicity, we depicted the classical trajectory with a solid line while the dashed lines stand for the quantum fuzziness).

And this, in turn, means that due to this slight but sometimes non-negligible degree of uncertainty, a quantum particle which was prepared with a kinetic energy of a value near to that necessary to jump over the wall but still classically insufficient, might yet possess, due to quantum uncertainty, that little bit of additional energy that will eventually allow it to overcome the barrier.

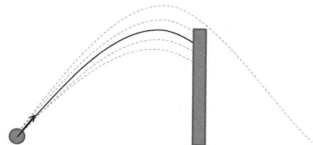

Fig. 97 The classical (solid line) and the quantum (dashed lines) possible trajectories of a particle overcoming a barrier.

Therefore, in QP it is always possible, though usually with low probability, that some particles will be able to overcome an obstacle even if they have a slightly lower kinetic energy than that which is classically necessary to overcome it. This classically not allowed 'slipping through' a barrier is called 'quantum tunneling' in QP.

This was a bit of a down-to-earth and intuitive way to put things. A more rigorous understanding of quantum tunneling can be achieved if we also recall the wave character of particles. With this perspective, we no longer conceive of a particle hitting a wall, or a layer of atoms, as a point-like or eventually extended well defined object which overcomes it or not, but rather we visualize it as a traveling probability wave which is in part reflected and in part transmitted through the barrier, just like a light ray can be partially reflected and transmitted by a glass plate. A wave can, due to diffraction effects, be deviated from classical paths, especially at the boundary of a sharp edge.

In microscopic structures it is more realistic to think of thin atomic or molecular layers. These too are 'barriers' that, for the sake of simplicity, can be approximated as the step function in Fig. 98.

If you insert the wavefunction into the Schrödinger's equation and constrain the particle appropriately with a potential V(x) which represents the barrier, then you will obtain a different solution for the wavefunction $\Psi(x)$ before, inside, and after the barrier (here $\Psi(x)$ is shown instead of its modulus square to show how the solution behaves inside and outside the well).

On one side (left of Fig. 98), an incoming and a reflected probability wave will overlap—that is, interfere—and result in the left standing wave (again, in principle a similar standing wave effect that we saw for orbitals or in the blackbody radiation cavity of Fig. 11).

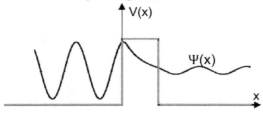

Fig. 98 The potential barrier (step function) and the wavefunction's tunneling.

Inside the barrier one gets an exponentially decaying wavefunction, which means an exponentially decaying probability with the thickness of the barrier, of finding the particle inside it. But for the transmitted wave (after the barrier), another solution will result: an oscillating non-zero solution, implying a non-zero probability to find the particle, having tunneled through the barrier. So, before the barrier, there are several regions where we will find the particle being reflected, since its kinetic energy was not sufficient to penetrate the barrier, but there is also a small probability to find it after the barrier as well. Statistically that means that multiple particles—even if all are prepared in the exactly same kinetic energy eigenstate—must have nevertheless slightly different energy values distributed over a mean energy due to the uncertainties that govern the quantum world. While most are reflected, a few will nevertheless make it through because these few will have a slightly higher energy content than the average value, which is considered the classical mechanical one (the right side of the graph of Fig. 94), and will be able to penetrate the potential barrier, even though they were prepared in a lower energetic state.

This is in stark contrast to the classical behavior where, if a number of particles all have the same energy, they will all behave together in the same way (they will all bounce back or all jump over the obstacle). Nobody has been seen to be able to walk through a wall! And yet, little tiny quantum particles are allowed to do that.

Atomic decays display quantum tunneling effects in a particularly striking manner. Quantum tunneling explains why radioactive elements decay. An atomic nucleus can be represented as a potential well where the protons and neutrons are trapped inside a square potential. A proton or neutron bound inside the atomic nucleus never has an entirely null energy: Because of Heisenberg's position-momentum uncertainty, it has a ground-state energy which in CM would be zero, but in QM is tiny but non-zero (recall how the same applies to the ground-state energy of an electron in the hydrogen atom). Due to this reason there is a tiny but non-zero probability to tunnel, since the particle will acquire sooner or later a sufficient impulse to overcome the nuclear barrier and traverse it.

'Sooner or later' means that the time at which the single decay will occur is completely undetermined, although one can speak of a mean lifetime for a nucleus. For special nuclear structural and energetic configurations, this period of time can be almost infinite (in that case the nucleus is considered stable), extremely long (eventually also longer than the life of the Universe), or also very short (say, of the order of a few microseconds). An element is considered radioactive when several nuclei decay every second. But the underlying principle is substantially the same. If we apply the Schrödinger's equation to the nuclear potential well and set the appropriate boundary conditions we obtain a solution for the wavefunction, which will furnish the probability to tunnel. Outside, there is a little wavy part of the solution (the small wavy left and right $\Psi(x)$ in Fig. 99) which is also non-zero at great distances from the atomic nucleus, and which means that there is a tiny but non-zero probability (given by $|\Psi(x)|^2$) to find a proton or neutron ejected outside the nucleus. Then a nuclear decay event has taken place.

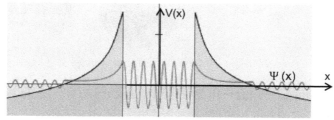

Fig. 99 Nuclear decay as a quantum tunneling effect.

An important point to keep in mind for all these examples is that there is no way to determine or decide *a priori* which particles will tunnel through and which will not. The tunnel effect, like all quantum effects, is a purely random stochastic effect. It is a probabilistic and completely unpredictable process which determines which of the single particles will tunnel through or not. You can never say *when* an element decays, only the mean amount of time that it takes for half of the nuclei to decay. This fact will be used later to illustrate the Schrödinger's cat paradox.

QT is not just something theoretical; quite the contrary, most of the physicists working on or with QT today are mostly concerned with all sorts of possible practical applications rather than questions about its foundations. One interesting example which shows how quantum effects are not only interesting phenomena for physical or philosophical ruminations but also can have very practical applications is the scanning tunneling microscope (STM). An STM is a microscope that uses quantum tunneling effects to investigate surfaces of materials at the resolution of the atomic scale. It was possible to build such microscopes only after the 1980s, because they require particularly precise mechanics, cryogenic techniques, and extremely fine positioning methods. Their working principle, however, is relatively easy to understand once you know about quantum tunneling.

Very briefly and schematically, Fig. 100 shows a tip, which is usually made of graphene.

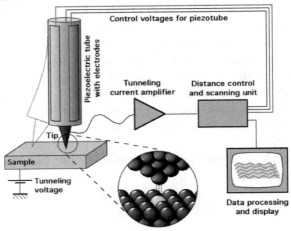

Fig. 100 The scanning tunneling microscope (STM).

This tip must be extremely thin and sharp, a sort of needle which is only few atoms thick itself, in order to position it exactly on each individual atom from time to time. This tip is positioned through another very precise positioning system made of piezoelectric crystals, the geometric properties of which can be varied and modulated very precisely by applying an electric tension, and which positions it above the surface that is to be analyzed. This graphene tip is positioned very closely to the atoms of the sample surface, onto which also an electric potential is applied. At some point they get so near that the electrons from one atom of the tip can quantum tunnel through the potential barrier induced by the tunneling voltage to the atom of the sample. This releases a tiny but measurable electric current which can be read out if appropriately amplified. The tip then moves along the sample surface by means of the piezoelectric positioning system and, through a computerized distance and position control, scans the area of interest step by step. It becomes then possible to reconstruct pixel by pixel an image that represents the atomic landscape. Thereby can one obtain images like that of Fig. 101, which is a graphene surface.

Fig. 101 Image of a lattice of graphite atoms as reconstructed by an STM.

In some sense we might say that STM allow us to 'see' atoms. But you should by now be able to understand that what these images actually show are probabilities. More precisely, the probability that at a specific position the microscope dip measures an interaction with the atoms. In what sense does this imply 'seeing'?

4. Zero-point energy, virtual particles and the Casimir effect

Another interesting aspect of QT is the so-called '*zero-point energy*' of empty space (also called the '*zero-point field*' or, more appropriately, the '*ground-state energy of the vacuum*'). According to CP, which, after all, reflects our intuitive understanding, empty space is just that: It is a space devoid of anything except empty space itself. If you empty a volume of all its molecules, atoms, and particles, and if you enclose this volume, say in a hermetic box, at the absolute zero temperature in order to empty it further of any thermal radiation (recall the blackbody cavity), and if you isolate this box from any external influence or perturbation from the environment, then we think of this volume as being perfectly 'empty', by definition. However, from a quantum perspective, things are not as simple as that. Even if we pump out all the air until the last particle and obtain perfect isolation at chilling temperatures, according to QM, that space is still not really empty (and the temperature still not exactly absolute zero).

We got a feeling in the preceding sections of some phenomena which have implicitly suggested this. We studied Heisenberg's uncertainty principle, which tells us something about a residual uncertainty when we try to determine a particle's position and momentum at the same time. Then there was also the time-energy uncertainty, which hints at the idea that, for short times, tiny extra energy amounts emerging from void space may be eventually possible. Then we said something about the quantum tunneling effect. A particle that cannot overcome a potential barrier classically can nevertheless do so in QP, because its wavefunction extends also beyond the barrier having non-zero probability to tunnel through it. All this together suggests, as odd it might seem, that there is always a small amount of energy literally appearing from 'nothing', and therefore it can do something that it classically would not be allowed to do.

Moreover, we studied the case of quantum superposition. Particles seem to be able to exist in a state of superposition, allowing them to be in more than one state at the same time. We found that it looks like a particle can be in several positions or can spin clock- and counter-clockwise at the same time. This assertion holds as long as we do not observe it, which means as long we do not perform a measurement and do not project the state vector onto one of its eigenvectors, that is, on one of its eigenstates. What then about the idea that, if particles can be in a superposition of states, why not also in superposition of different energy states? From these considerations one can shift the reasoning also onto empty space, or what is believed to be classically 'empty'. Energy from nothing? Could that be true?

Frequently you will see, among textbooks or physicists talking popular science to non-academic audiences, an approach to this question framed by invocation of Heisenberg's principle or the energy-time of uncertainty. Sometimes, especially in the popular literature, one reads about the time-energy uncertainty as a direct consequence of energy fluctuations in empty space. In this scenario one thinks of empty space filled with a continuous fluctuating quantum energy foam which is allowed to violate for very short time intervals energy conservation according to the time-energy uncertainty. If we conceive of sufficiently small time lapses Δt—so goes the theory—the energy is indeterminate according to the time-energy relation, and at microscopic scales (for space intervals not larger than light can travel in that time interval, that is $\Delta x = c \cdot \Delta t$) we can imagine energy popping into existence for a very brief moment and disappearing again.

In some sense this might sound very convincing and could be taken as a first approach of understanding the nature of empty space in QP. After all, where does that ΔE in the time-energy uncertainty relation come from? We expressly excluded any classical measurement error due to imperfect measurement devices. This little bit of energy that comes into or goes out of existence remains something quite mysterious if we recall that the energy conservation law is one of the most universally proven facts of CP. According to this law, energy can be transformed (from heat into mechanical energy, chemical into electrical energy, etc.), but can never be created or destroyed. Does QP violate the principle of energy conservation?

A possible way out of this is to admit that the energy conservation law can be violated for an extremely short time (we have seen that a typical atomic transition lasts for about fractions of billionths of a second), but then the energy budget must be, so to speak, 'restored'. In other words, in QM, tiny bits of energy are allowed to pop in and out of nothingness if they exist only for a time lapse which is almost an instant. One also finds professional physicists talking, somewhat improperly, about a "borrowed energy from the vacuum", calling it 'zero-point energy'. This is how the law of energy conservation can be saved on the classical and macroscopic level, as we perceive a world in which the shortest time of which we can be aware is no shorter than a tenth of a second or so.

While we can view things happening intuitively according to this model, a more rigorous analysis shows that this 'extra energy' is conceived of as a ground state of the vacuum and is obtained by the non-commutative property of observables, *not* through the time-energy uncertainty relation. Because, as we pointed out in one of the previous sections, there is no time operator, contrary to the position and momentum operators, the zero-point energy is not a consequence of the time-energy uncertainty relation. The vacuum state is considered the lowest possible quantum state of empty space and is obtained by a mathematical theory of QM, which goes by the name of 'second quantization'. The first quantization is the one we intuitively introduce here: the introduction of operators for observables. The second quantization expands these again as a sum of more fundamental operators (called 'annihilation operators' and 'creation operators')—that is, a mathematical procedure which quantizes every point in space in terms of harmonic oscillators and which is able to correctly describe special relativistic quantum mechanical phenomena (though not, or still not, general relativistic ones). It is from this that physicists obtain, from their calculations, a non-zero ground state energy of the vacuum, and which can be imagined, somewhat inappropriately, as a quantum foam in which every space-time point 'fluctuates'. In this interpretation, we do not perceive these fluctuations at large spatial scales because they become relevant only for small microscopic regions. Just as a small rippling on the surface of an apparently perfectly flat object might be invisible to the eye, if you were to look at it through a microscope, you might discover it to be very rough instead.

In fact, if we continue zooming into space regions as small as 10^{-34}m, which is an incredibly small scale (a zero followed by 33 zeros after the comma until you reach the first non-zero digit) called the 'Planck scale' (Planck was the first to describe it), empty space is considered to be full of these energy fluctuations. This is often depicted as a space-time foam, a sort of 'white noise' pervading empty space at very small scales, as shown in the last magnification on the right-hand side of Fig. 102.

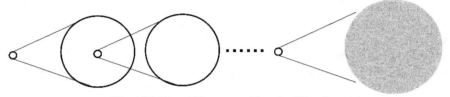

Fig. 102 Magnifying space till to the 'Planck scale'.

At these spatial scales, space is no longer empty, but a place where any sort of extremely fast bubbling energetic events in extremely short time spans occur (about 10^{-44} seconds, called '*Planck time*'). So, the energy state of the vacuum in QM is not zero. The zero-point energy, despite what the somewhat-misleading name might suggest, is not a point at which the energy is zero but, rather, a non-zero energy ground state of the vacuum.

A word of caution about the concept of 'quantum fluctuation'. You might frequently hear the general public—and often physicists, too—speaking of these quantum fluctuations of the vacuum at small scales. We have done this here too, pictorially. This interpretation might be useful for an intuitive understanding, but it isn't entirely correct. A more rigorous way to interpret the quantum mechanical vacuum is not that of describing it with fluctuating energy states, but quantum superpositions. Nothing is 'fluctuating', but the vacuum is in a superposition of possible energy states, and only one of these actualizes when observed or, as we shall see now, when particles interact. To clarify the difference between the naive and misleading idea of 'vacuum fluctuations', think of the 50 percent probability of observing the spin-up or spin-down state of a particle's spin, as we have seen with the SG-experiments. We have seen that in QM, it is entirely possible that a particle can be in a superposition of spin states, and only when we measure it—that is, only when a spin-up or spin-down state is realized—will the measurement project the state vector onto one of the possible eigenstates. But before the measurement, it would be misleading to conceive of the particle as 'fluctuating' or 'oscillating' between the two spin states. There are no fluctuations of states. There is a superposition of states and the subsequent realization of one of the potential events by a measurement—a very different thing.

We should conceive of empty space in the same way. At a microscopic level and for very short time intervals, empty space is in a superposition of energy states. On the other hand, what really fluctuates are the outcomes we obtain at each measurement. If we perform several (extremely fast) measurements looking at empty space, we would obtain a set of fluctuating data (again, this has nothing to do with the imperfections of the measurement device). This is another important example of how, in QP, counterfactual definiteness must be applied with great care. The fact that measurement results fluctuate does not imply that the system we are measuring fluctuates in time from one state to another.

Another physical consequence from empty space having a zero-point energy, that is a non-zero ground state energy, is that it will never be possible, even not in principle, to cool down a material object to the theoretical absolute zero temperature (0 K or -273.15 C°). A system completely devoid of any EM radiation must still possess a quantum mechanical zero-point energy, and this in turn provides the tiny but non-zero kinetic energy to the atoms or molecules it is made of. Since, per definition, temperature is the measure of the average kinetic energy of the particles making up a body. That's why scientist were able to build cryogenic chambers that are able to create an almost perfect vacuum at chilling mK, eventually μK or even nK temperatures, but will never be able to reach the absolute zero temperature of exactly 0 K. The fundamental laws of QP will forever prevent us to realized this ideal classical thermodynamic state.

And this leads us to the next step: that of the conception of '*virtual particles*'. In this scenario, one also conceives of the existence of virtual particles filling empty space and popping in and out of existence. In fact, if empty space can be filled for short times in small, microscopic regions with non-zero energy states, we can also think of the vacuum as being filled with particles having that energy, as photons, or even with a mass, according to the famous Einstein mass-energy relation, which states an equivalence between mass and energy as $E = m\,c^2$. Photons, or even massive particles like electrons or protons, are allowed to come into existence from empty space and, almost instantly, disappear. This might seem a bit too farfetched of an idea, but the fact is, once accepted, the whole quantum field theory (QFT) resulting from this assumption is just that which turned out to be the right one and which was tested experimentally many times.

To clarify the meaning of virtual particles in QM, think of two approaching particles as two electrons in Fig. 103. It is one of the typical diagrams, called '*Feynman diagrams*', that we will discuss in more detail later. It is a different way of conceiving of force fields—no longer as continuous regions surrounding a particle or a body, like the gravitational or electric field of CP, but as particles that exchange other force particles. This is one of the main aspects that distinguish QFTs from classical ones. Fields are quantized. Here, when the electrons approach and come sufficiently close to each other, the region of space that separates them—the '*vertex region*'—is so small that due to the vacuum energy, a certain amount of energy becomes available, though only for a very short period of time.

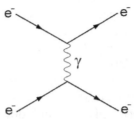

Fig. 103 Virtual photon exchange in the scattering process of two electrons.

During this brief time interval, the zero-point energy manifests in the form of a virtual particle, here a virtual photon, which is exchanged between the two electrons, resulting in an impulse exchange, deviating them from each other. (They have the same charge and therefore must repel each other—that is, we are talking about a scattering process.) The time-energy uncertainty tells us that the distance between the two electrons must be small enough that the photon can travel from one electron to the other in a shorter time than that allowed by the time-energy uncertainty. Then, at the end of this force-exchange process, the photon must be absorbed—that is, it ceases to exist.

Virtual particles can never be observed directly and do not exist in the same way that other particles do. We might say that they pop in and out of existence for that short time period and in that tiny space volume to allow for a force and momentum exchange, for the scattering process to occur; however, they cannot be compared to other 'real' particles that can be observed directly. That is why, after all, they are called 'virtual particles'. Things are a bit more intricate, and we will see this in more in detail when we talk about Feynman path integrals and diagrams. However, for the moment, this might be an image that can help our intuitive understanding of things.

The question is: Do these visualizations represent a real process? Do virtual particles really exist as in the Feynman diagrams? The fact is, his approach was extraordinarily successful in describing particle physics with a theory that today goes by the name of the *standard model of particle physics* based on Feynman's *quantum electrodynamics* (QED) and was confirmed by innumerable experiments in particle accelerators. There can be no doubt that it correctly calculates the phenomena we observe. However, we must be aware of the fact that Feynman diagrams represent only an intuitive tool for our limited human mind to depict interactions, the complexity of which escapes our cognitive abilities. They are an attempt to force into the realm of the microcosm our classical representation of objects in the realm of the macrocosm.

Another proof regarding how a vacuum can manifest its quantum superposition state, which reveals itself as ripples of energy over a large space volume, comes, paradoxically, from a completely different realm, one that is the opposite of the microscopic world: cosmology. According to modern big bang theories, all the matter and energy of the Universe was once concentrated in such a tiny volume that quantum mechanical effects were very relevant during its first instants of birth. Even if all the matter were distributed in a perfectly and ideally homogeneous manner, the quantum effects of the vacuum must have caused tiny inhomogeneities that should be observable in today's microwave cosmic background radiation. In fact, this is what we already highlighted in the section on the blackbody radiation. Inhomogeneities of the cosmic microwave background are observed (see Fig. 17). Though the ripples of inhomogeneity of the cosmic background are very tiny (only a few thousands of degrees over a 2.7 K temperature), these could be confirmed by measurements from satellites in space.

Particle physicists instead take as proof of the existence of the zero-point vacuum energy and the existence of virtual particles an experiment performed by Dutch physicist Hendrik Casimir (1909-2000) in 1948.

In the '*Casimir effect*', the ground state of the quantized electromagnetic field, that is, the EM vacuum energy, which can also be seen as mediated by virtual photons popping in and out of existence, is thought to be responsible for the attraction between two almost perfectly flat, parallel and uncharged metal plates distanced from each other by only a few micrometers (see Fig. 104).

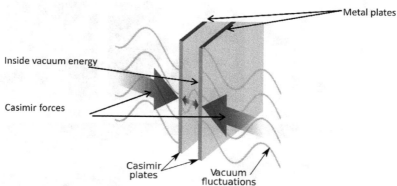

Fig. 104 The metal plates experiment confirming the Casimir effect.

According to the standard interpretation, the Casimir forces arise because the ground state of the vacuum between the plates will have a different EM modal configuration than that of the outside vacuum, as its ground states are constrained by the two metal walls (sort of the modes we considered in the blackbody cavity). This implies that the virtual photons inside the plates have a different energy than those outside it. This leads to a difference in the energy density between the two spatial regions inside and outside the plates and produces a force pulling the plates together: the *'Casimir force'*. (The variation of energy in space does always produce a force.)

This effect has been repeatedly tested with increasing precision since then, confirming the theory that predicted it. Therefore, one hears several physicists referring to the Casimir effect as evidence of a zero-point energy, that is, a foam of virtual particles, supposedly filling empty space. However, some physicists do not share this view, despite the popularity of this interpretation. In fact, Casimir himself originally described the same phenomenon as an effect of *van der Waal's forces*, the intermolecular forces responsible for the attractive or repulsive forces between polarizable molecules at short distances. Casimir forces can be described as an interaction between the materials. The reason why, in QP, the zero-point energy interpretation prevailed is that with this theory and for elementary quantum particles or structures one encounters, the mathematical calculations are easier to perform. This is not because it adds something new to the physical picture. In chemistry and biology, however, scientists prefer to speak of van der Waal's forces.

At any rate, however one interprets the Casimir forces, the most amazing and unexpected display of this effect in Nature comes from the animal kingdom: from the tiny and elegant lizards called 'geckos'.

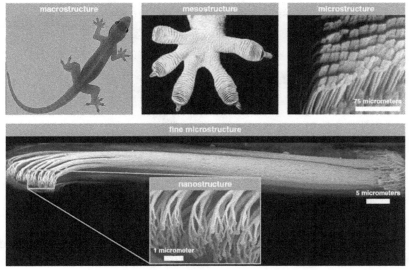

Fig. 105 Micrometer- and nanometer-scale view of a gecko's toe.

Most members of this species are quite skillful at climbing walls or any kind of flat surfaces (like, for instance, a vertical glass window) without using liquids or resorting to absorbency forces or surface tension. This ability comes from their adhesive toe pads, which enable attractive van der Waals' forces between the setae structures (that is, the stiff, hair-like or bristle-like structures) and the surface. These bristles are approximately 5 μm in diameter. (The smallest human hair is about 15-20 μm.) The use of the small van der Waals attraction force requires very large surface areas: Every square millimeter of a gecko's footpad contains about 14,000 hair-like setae. The end of each seta consists of approximately 1,000 'spatulae' shaped like an isosceles triangle. The spatulae are approximately 200 nm on one side and 10–30 nm on the other two sides. The structures are invisible because they are below the wavelength of visible light. This plethora of spatulae all act like tiny Casimir plates, which together sum up to an adhesive force allowing the gecko to climb walls without further aid.

5. The Bohr-Einstein debate and the Copenhagen interpretation

If you have followed so far, you will probably agree that the picture of the Universe that emerges from QP is really weird, at least from the point of view of our everyday human experience. Many strange things happen. We have the wave-particle duality whereby, before we observe it, particles do not behave as particles at all but as waves, and yet, when we measure it, we always get only point-like interactions as an answer. There is the Heisenberg's and the time-energy uncertainty principle in which these particles seem to have intrinsically fuzzy properties in their position, momentum and energy, so that we can no longer conceive of them as objects with definite physical attributes but only as objects smeared out in time and space. To describe all of this, physicists have come up with what they called the 'wavefunction', which has that annoying property of collapsing immediately at the instant when an observation occurs, and when nobody knows for sure whether it is a thing that has some reality 'out there' or whether it is simply a mathematical abstraction. We saw then that particles can tunnel through a barrier, even though they do not have the classically required amount of energy to do so. Then, because of the ground state energy of the vacuum, empty space seems to be not so empty after all and, according to a probably bit too naive interpretation, lots of particles can quickly pop in and out of existence. Moreover, events in the quantum domain must be described with probabilities because everything is random and unpredictable. This is not the case in the sense of CP, in which the description of unpredictable events with statistics reflects our ignorance and limitations in computation. In QM, events can occur entirely without a cause; it is a 'cause-less' realm. How is it possible that in the double slit experiment, by maintaining exactly the same experimental conditions and eliminating all possible measurement errors, nevertheless, one electron hits the screen in one position, while another electron will be found in a completely different position? This is only the beginning of a long list of weird things in QP.

What emerges from this picture is a reality that looks tremendously counterintuitive to our understanding of the physical world. From childhood, our brains have been trained to see and interpret phenomena in terms of CM, in which objects have well-defined properties and do not have these ambiguous behaviors that QP displays. On the other hand, it is a fact that QP, as it is, with this mathematical structure founded on the notion of the wavefunction or the state vector, which delivers probability waves, works exceptionally well. It is able to describe the phenomena from particle to molecular physics and it makes precise quantitative predictions where CP fails.

Yet, while from a mathematical perspective everything works fine, from a more philosophical one which tries to make sense of all this and wants to satisfy the human thirst to know the deeper meaning and essence of things, even the brightest minds were not able to find a clear description. Math works fine, but at the epistemological and ontological level, nobody agrees. We also will see how a great physicist and genius like Schrödinger was puzzled by the implications of QP, which led him to propose the famous quantum cat paradox. However, among the physicists of the time, the one who was most dissatisfied with this state of affairs was Albert Einstein. This was because Einstein's own theories, above all special and GR, relied heavily on the opposite world conception in which the focus is rather a mathematical approach that preserved elements of the classical mechanics of Newton and Galileo. In relativity, everything has well-defined properties, particles follow precise paths and, at least in principle, everything is predictable without statistical considerations, just as in CP. Contrary to QP, relativity is a purely deterministic theory. That is one of the reasons why relativity is sometimes categorized as still being a 'classical' theory, though it superseded Newtonian mechanics. Moreover, as we will see in the section about quantum entanglement, QT seems to suggest even FTL interactions, something which Einstein could certainly not take seriously because his own theory of relativity strictly forbade it. His approach to physical reality was so successful that Einstein was not willing to accept that QP is a complete theory describing a real world. He looked at it as only a provisional model that, sooner or later, had to be replaced by a better theory to account for all those strange and counterintuitive facts. In this context, Einstein once formulated his belief that God does not play dice, referring to that mysterious randomicity of QP. He expressed this in a letter to Born in 1926: *"Quantum mechanics is certainly imposing. But an inner voice tells me that it is not yet the real thing. The theory says a lot, but does not really bring us closer to the secret of the 'Old One.' I, at any rate, am convinced that He is not playing at dice."* [8] Much later, in 1944, he returned back to this point: *"We have become Antipodean in our scientific expectations. You believe in the God who plays dice, and I in complete law and order in a world which objectively exists, and which I, in a wildly speculative way, am trying to capture. I firmly believe, but I hope that someone will discover a more realistic way, or rather a more tangible basis than it has been my lot to find. Even the great initial success of the quantum theory does not make me believe in the fundamental dice-game, although I am well aware that our younger colleagues interpret this as a consequence of senility. No doubt the day will come when we will see whose instinctive attitude was the correct one."* [9]

On the other side of the front was Niels Bohr. Bohr was more prone to accepting the strange reality that QP suggests. This is not surprising because he had a rather different professional experience. His theory of the atom structure was not as successful as Einstein's relativity and was quickly replaced by the quantum mechanical picture itself. Bohr's success relied more on discoveries which had their roots in QP rather than in CM. Therefore, while Einstein was puzzling on the nature and essence of QP, Bohr took a more pragmatic view of the whole problem. For him, it was easier to embrace QP as a matter of fact that we should accept as it is. From these two diverging points of

Fig. 106 Niels Bohr and Albert Einstein in 1925.

view emerged a debate over the conceptual foundations of QP, which nowadays is known as the '*Bohr-Einstein debate*', and which took place from about the 1930s until the end of WWII.

Einstein's approach was to look at loopholes in the theory, some aspects in the current foundations, which must lead to contradictions or inconsistencies. He thought that such flaws could finally vindicate the classical understanding of physical reality. Because he found it unacceptable to consider a theory that displays such a fuzzy and undetermined realm as QP does, he tried hard to find some way out, something that would show us that the theory is incomplete and requires further improvement. By 'incomplete', it is meant that the theory must have *'hidden variables'*, which assumes that the quantum events are caused by something we haven't yet discovered and which would explain the fuzziness, uncertainty and apparent a-causality of QM.

Einstein and other physicists hoped to explain quantum indeterminism and where it appears by resorting to a deterministic conception of reality, just like the theory of gasses explains the molecular chaos of molecules and atoms with the Brownian motion, that is, with the excitation of molecules due to thermal effects. Analogously, a future refined QT should explain its indeterminacy as a superficial manifestation of hidden components, or at least more generally unknown variables, which in the current, supposedly incomplete form of the theory are still not evident. In fact, it is difficult to escape the impression that in a world where things happen only by chance, just out of the blue, apparently with no causal relation, we are eventually missing something. The very principle of cause and effect seems to no longer work here. Is there perhaps something we still don't know, some hidden variables in the theory we don't see, that, once found, will explain all this reality that is so alien to our daily common sense by returning to our good old Newtonian and deterministic understanding?

Bohr instead argued that there are no hidden variables, QM is a complete theory and we should accept it as it is, because, after all, it works well, it is mathematically consistent and it furnishes the quantitative predictions we need. This is precisely what science should be about. The rest is philosophy that should not bother physicists.

For the rest of his life, Einstein tried to reconcile QM with a deterministic understanding of the world. He published articles in which he proposed interesting and very smart experiments to show why QM couldn't be a complete theory. The most famous one was written with Boris Podolsky and Nathan Rosen, and it illustrated the so-called 'EPR paradox' that we will discuss later. On his side, Bohr answered Einstein's attacks on the foundations of QT with other articles in which he showed Einstein's fallacious thinking. It was an interesting battle between two titans of science. To provide an example of how this debate developed, it might be interesting to look at how Einstein argued and Bohr answered.

For instance, Einstein proposed a thought experiment to suggest that QM cannot be a complete theory for the following reasons. Let us send a single particle through a single slit device as shown in Fig. 107. (In truth, Einstein argued initially with a double slit, but for the sake of simplicity let us illustrate his ideas in a manner that Bohr summed up later.)

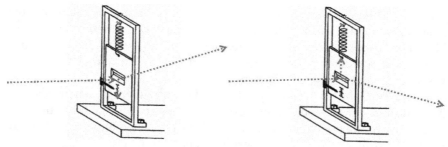

Fig. 107 The single slit thought experiment of Einstein vs. Bohr

This single slit hangs at a spring, as depicted in Fig. 107, which Bohr himself drew. (Note how he added the screws; this might say a lot about Bohr's personality.) As we know, when the particle passes the slit, it will statistically suffer a deflection from the straight incoming direction line according to Heisenberg's uncertainty principle or, equivalently, if you prefer, because of the wave-particle duality. However, if we maintain that the conservation of momenta is a universally true law, this deflection of the particle by the slit must also cause a small recoil of the slit itself in the opposite direction. We should be able to measure this recoil by observing how the slit, which is directly attached to the spring, must move and eventually cause an oscillatory movement of the spring and slit structure. If the particle is deflected upwards (Fig. 107 left), the slit will suffer a recoil downwards. On the other hand, if the particle is deflected downwards, the slit will suffer a recoil upwards (Fig. 107 right). Once detected on the other side of the device, we know how large this deflection was. This means that, at least in principle, one should be able to trace back the position of the particle when it traverses the slit (the position of the slit itself) and also its momentum at the same time by observing the slit recoil. However, this is contrary to all that Heisenberg's uncertainty principle states, which excludes the notion that we can know the position and momentum of a particle at the same time, even in principle.

This is, of course, only an ideal representation of the experiment. It would be difficult to build such devices for tiny elementary particles, as we would need a perfectly friction-free slit and a spring system which is macroscopic compared to the particles; extremely precise measurement devices would be necessary to detect any motion produced in it. That is the reason why they are called *'thought experiments'* (from German 'Gedankenexperimente'); they cannot be realized practically but are nevertheless logically and physically consistent. This kind of thought experiment was, for Einstein, a perfectly legitimate experimental arrangement that contradicts the conventional interpretations of QM—at least Heisenberg's uncertainty principle, which for this reason must be considered not as fundamental but a manifestation of a deeper hidden variable theory, which in turn means that QP must be considered incomplete. This was how Einstein's argument went—an argument that seemed quite convincing to Einstein, who certainly was a very smart guy.

However, Bohr was a smart guy too, and he disagreed. In fact, he showed that Einstein forgot something very important in this picture: that the uncertainty of the position of the slit itself must be taken into account. Also, for the slit structure as a whole, despite its being a macroscopic object, we must apply the uncertainty principle. The point is, the tiny recoil of the slit— say, caused by an electron, for instance—would be so extremely small that we must use the rules of QP to determine it. In his response to Einstein, Bohr showed that if we also take into account Heisenberg's uncertainty relations for the slit, we must also conceive of its position as slightly, but not negligibly, uncertain. The oscillations of the slit would be so small, its position and the momentum shifts due to the electron's recoil would be of the same order of magnitude as the position and momentum uncertainties of the particle that traverses it. This uncertainty over the slit position itself is just that uncertainty that will remove the information about the magnitude of the slits' recoil and, ultimately render Einstein's thought experiment useless in determining the particle's momentum.

Therefore, Einstein's thought experiment relied on a fallacious thought. It will not allow us to determine, at the same time, the position and momentum of particles, not even in principle, and he could no longer state that this kind of reasoning leads to any demonstration of the incompleteness of QM. In essence, what Einstein seemed to have forgotten was that there is no reason to believe that Heisenberg's uncertainty principle no longer holds for macroscopic objects. The uncertainty relations always hold for any microscopic particle to any macroscopic object—even for things as big as a planet. However, at these scales, the quantum effects are so extremely small and undetectable, there is no need to take them into account. Already, the quantum uncertainty on position and momentum for a tiny object like a grain of sand is far beyond any possibility of detection, even for our most advanced measurement apparatus. We can simply forget about it. Not so, however, for thought experiments like that with the recoiling slit. It might help young students, those who are struggling to learn the basics of physics, to keep in mind that great scientists like Einstein also made quite blatant errors.

This was one of the most famous examples illustrating the Bohr-Einstein debate. Bohr and Einstein engaged each other in a friendly way, as gentlemen, with respect and mutual admiration (regretfully, something one rarely sees nowadays) but defended their own views with papers and articles.

Einstein refused, until the last of his days, to give up a classical understanding of the world. We will see how modern experiments, which were performed after Einstein's and Bohr's deaths, clearly indicate that a deterministic local hidden variable theory is no longer tenable, at least not in a locally realistic form that Einstein would have desired.

On the other hand, we might also question whether Bohr's pragmatic approach is the final answer to QP. Bohr's approach was much more at ease in this intellectual challenge. While Einstein had to think of more or less complicated thought experiments trying to falsify the current conceptions, Bohr essentially had to do nothing except reply to Einstein's attacks and defend the status quo. Bohr's attitude towards Einstein's doubts and his interpretation of QP is summed up in the so-called *'Copenhagen interpretation'* of QM—an interpretation which, since then, was adopted by almost the entire community of physicists for a long time to come. It was considered the final nail in the coffin on the subject and is still considered as such by many physicists to the present day. However, in the last three decades, thanks to new technological advances that allow us to perform experiments previously impossible to perform and due to the well-funded research which attempts to build quantum computers, the debate has returned to the stage and continues to occupy several philosophers of science and some physicists. We will go deeper into this subject in the following chapters. However, it is important to know what, essentially, the Copenhagen interpretation of QM is about so that we have an idea and perception of the kinds of philosophical and ideological backgrounds that influence QP to this day.

In reality, the Copenhagen interpretation is not an official declaration that was announced by someone or during a congress or any scientific institution. It is an interpretation which began to arise during the late 1920s and early 1930s and is associated with Bohr, who was a professor at the university in Copenhagen, but there is no piece of paper or historical act that clearly defines it. It loosely expresses an understanding, an interpretation of things and a spirit of the time. That is why the author feels authorized to express it in his own words, though there are several other ways to express it as well.

One way to state the Copenhagen interpretation is simply to say that: *"Quantum mechanics is about measurement outcomes; we should not attempt to infer objective realities from it."* This is a strictly pragmatic and utilitarian approach. Stop thinking about the deeper philosophical issues, do your calculations, see if they fit with the experiments and be happy with that, especially if they do fit. In fact, this approach has its logic: The human and the scientific mind is one of senses, a mind which evaluates, infers and extrapolates an interpretation of the world from empiric data, which, however, can lead to different models and interpretations of what is seen.

Think, for example, of the Gestalt figures. You might see two faces or a vase, but the figure remains the same. So, in this view, it is dangerous to make these extrapolations. According to the Copenhagen interpretation, we must stop there, at least as physicists.

Another typical statement that could be ascribed to the Copenhagen interpretation is of the kind: *"The wavefunction is only a mathematical abstraction which furnishes probabilities that a specific event occurs. It is not a real physical object."* This, again, is a mathematically pragmatic approach. Simply do your calculations with the wavefunction, but do not attempt to infer from that a deeper ontology of things, that is, the nature of how things are in reality independent from what we perceive. After all, what is 'reality'? What follows from that also is that, according to the Copenhagen interpretation, *"there is no such thing as a wavefunction collapse"*. This should be considered only a mathematical update of the information we have about the state of a system. What collapses is the lack of information about the system's state. Before the measurement we don't know, while after the measurement we know something more about the state of a particle or a system. So, in this view, it is a quite natural thing that a collapse seems to occur; it is just an abstract formal act of encoding information from one instant before the measurement to that after the measurement. The problem is solved because, according to this standpoint, no problem exists in the first place.

Yet another statement of the same point of view: *"The wavefunction describes the state of a system and is all that can be known about."* The wavefunction is all that can be said; it is futile to ascribe a deeper meaning and reality to it, just as it is futile to attempt to get more out of it.

Finally comes the question of why, then, this discrepancy exists between the physics we observe in the microscopic vs. the macroscopic realm. Why do we see things so differently in the classical world as compared to the quantum one? The Copenhagen interpretation answers with the well-known '*Bohr's correspondence principle*', which can be summed up by saying that *"a large collection of quantum particles will exhibit classical behavior with a growing number."* As the number of particles grows, even if they have very different properties and states, these properties average out when a certain quantity of particles is present, and just that average emerges on the macroscopic scale, defining the qualities of the world we know.

In other words, according to those who embrace the Copenhagen interpretation, there isn't much to understand and investigate in terms of the deeper meaning of things. Simply let us accept what we observe and posit it as a principle a priori and work with it. Bohr summed this up as follows: *"Our task is not to penetrate into the essence of things, the meaning of which we don't know anyway, but rather to develop concepts which allow us to talk in a productive way about phenomena in nature."* [10] This attitude was not new, it has its roots in the very beginnings of the history of science. When Newton was asked for the reason for the properties of gravity he answered with his famous *"hypotheses non fingo"*: *"I do not frame hypotheses. For whatever is not deduced from the phenomena must be called a hypothesis; and hypotheses, whether metaphysical or physical, or based on occult qualities, or mechanical, have no place in experimental philosophy."* [11]

This scientific utilitarianism which is especially incarnated in the Copenhagen interpretation might sound a bit dry and not really satisfying for those inclined towards a deeper understanding of things—an inclination that seeks to find the ultimate realities. However, it is a fact that it worked well for several generations of physicists. This pragmatic approach gave us the modern standard model of particle physics, which was one of the most successful theories not only of physics but of perhaps all science. It also gave us the modern theories of particle astrophysics and cosmology, as well as allowed us to build many devices, especially in the electronics branch and in information technology. Therefore, contrary to popular belief, for a long time, apart from rare exceptions, physicists no longer worked on the conceptual foundations of QP, but were busy developing quite complicated mathematical and abstract theories that allowed them to make calculations predicting the empiric data. Not much more. Even today, most are not working on it, as it is highly unlikely that they will find a job and make money from this. A now quite famous sentence with which N. David Mermin, a physics professor at Cornell University, summed up the Copenhagen interpretation and its attitude towards foundational issues is: *"shut up and calculate"*. This is how physics was, and still largely is, taught in universities and, regretfully, in schools. Don't ask questions; just do your exercises and your homework, which are complicated and time-consuming enough and which leave almost no time (and psychological energy) to do anything else. No wonder that, in a world so profoundly conditioned by science and technology, the younger generations are less inclined to apply for studies and jobs that require technical skills.

To sum up with the words of philosopher of science Tim Maudlin:

„Nowadays, the term [Copenhagen interpretation] is often used as shorthand for a general instrumentalism that treats the mathematical apparatus of the theory as merely a predictive device, uncommitted to any ontology or dynamics at all." – "Sometimes, accepting the Copenhagen interpretation is understood as the decision simply to use the quantum recipe without further question: Shut up and calculate." [12]

However, while the human spirit of enquiry can fall asleep for some time (and on some occasions for a very long time), it cannot be suffocated forever. As we will see, in the late 1960s, Sir John Stewart Bell, a Northern Irish physicist, came along with very interesting considerations that re-opened the gates to new speculations and new thought experiments, which could historically be considered a continuation of the Bohr-Einstein debate. What also contributed largely to this quantum renaissance was the fact that nowadays we have technologies that allow us to build real experiments, in the laboratory, of what previously could, at best, be only thought experiments. Moreover, the attempt to create technologies of quantum cryptography for secure transactions and to build quantum computers—that is, computers which no longer work with traditional electronic gates but instead with logical gates based on quantum logic and quantum effects—has initiated, in the last two decades, a huge research effort in the field of quantum information. This inevitably forced us to reconsider some aspects of the foundations of QM and led to a renewed interest in philosophical questions which had been laid to rest for about the last thirty or forty years.

V. Quantum entanglement

1. Quantum scattering and indistinguishability

Before we proceed to the famous phenomenon of quantum entanglement, to better understand what it is about, we must introduce the concept of *'quantum indistinguishability,'* which emerges from quantum scattering processes. This is another weird thing about QP. Quantum indistinguishability might at first look like the classical indistinguishability of two objects having the same physical properties between which one cannot distinguish. This might not sound particularly exciting. Sometimes it is, indeed, presented by physicists in this way—physicists who do not bother about the subtler conceptual foundations to a popular audience or to their students. However, a profound and important difference exists between classical and quantum indistinguishability.

For our everyday experience, every physical object is distinguishable, even inside a collection made of many objects having exactly the same properties. When, in our human macroscopic classical world, we want to maintain two equal objects as distinguishable (say, subjected to a complicated reshuffling process that is difficult to keep track of), we can do this simply by labeling them, say, with label A and label B. Then start the process—for example, shake them inside a box and later look at where one or the other object is simply by reading the labels.

One might argue that appending labels is already differentiating between two unequal objects. Then, alternatively, we can follow their paths separately, looking at where one or the other object moves in space and time, even without labels. If this will be too fast and complicated, we can eventually record it using a video camera. For example, consider two billiard balls having exactly the same properties (mass, size, color, etc.) and colliding. This is our everyday experience with the classical mechanical scattering process, which involves hard and well-defined objects, just chunks of matter. (This collision process could be elastic or inelastic, but that fact is not so relevant to our considerations here.) If they interact, we can nevertheless distinguish one from the other, for example, by letting them collide head-on, as in Fig. 108, and then observing that the first is scattered in one direction while the second is scattered in the other direction.

We can track their trajectory simply by looking down at the billiard table from above. In the classical view, we determine, during the process, whether billiard A went along path i and billiard B went along path j, or whether, to the contrary, B went along path i, while A went along path j. It is as simple as that.

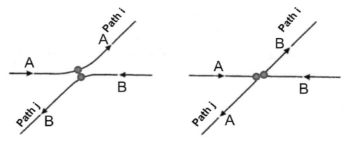

Fig. 108 Classical mechanical scattering process.

So, in this Newtonian mechanical perspective, nothing is fundamentally indistinguishable, and if it is, that is only because we have not looked at the process carefully enough; in principle, distinguishability is always possible. Classical indistinguishability is only a notion which reflects our ignorance; it is not an ontological statement.

If, however, we would like to know what happens during a scattering process between two elementary particles, say, two electrons, we must consider the collision as not like that of two billiard balls coming in direct contact with each other (that is, 'touching' and repelling each other); rather, we must take into account the fact that an electric field surrounds them.

In this case, because the electrons have the same negative electric charge with the same polarity, they repel each other. For tiny charged particles, the scattering never involves direct contact between particles; the repulsion occurs because of the mutual interaction of the extended electric field of both particles. Therefore, one must refine the previous mechanical scattering model with that of Fig. 109.

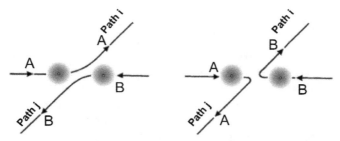

Fig. 109 Classical electric scattering process.

The trajectories are deviation paths caused by the mutual interaction of the two electric fields, without the particles having any direct contact, at least not in the classical mechanical sense of which we are accustomed to thinking. Keep in mind that also, at the macroscopic scale, there is no such thing as 'direct contact', as naively suggested by Fig. 108. Every interaction, 'contact' and scattering between material bodies—including that of billiard balls—is microscopically an electric repulsion between the electron clouds of the atoms and the molecules of which a body is made. In our everyday lives, we perceive the existence of hard objects like stones, but this 'hardness' boils down to microscopic interactions between particles of the kind in Fig. 109. Also, things are a bit more complicated than that, as we shall see later.

Therefore, as to what regards distinguishability, things turn out to be a bit more complicated. You may have already seen another naive assumption we used for the case of the billiard balls. To observe the collision process, we must have some light in the room, which the billiard balls reflect, to obtain an image that tells us which went along which path. At human macroscopic scales, the reflection of photons on billiard balls is such a tiny perturbation of the system that we can consider it completely negligible. For two electrons, however, we can't forget this little photonic perturbation. Here we are in the domain of Heisenberg's microscope. In the microscopic realm of elementary particles, the light photons would scatter the particles we would like to distinguish and would disturb their trajectories, rendering impossible the kind of classical distinction process mentioned above. However, as we have already mentioned, the perturbation argument is not entirely correct, or at least it isn't the whole story that Heisenberg's uncertainty principle is telling us. As we will see later, it is possible to build experimental setups in which the perturbation is negligible or even absent and nevertheless the uncertainty principle remains unavoidable.

Let us, therefore, consider a more quantum mechanical version of the whole problem. Imagine again that the two identical particles—say, two electrons, A and B—scatter with each other, as in *Fig. 110*.

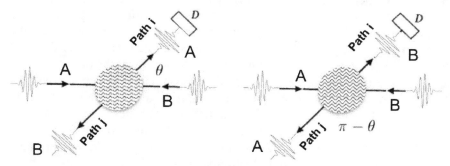

Fig. 110 Naive quantum scattering process.

We do not shine any light particles on them to track their whereabouts, so there is no perturbation. Rather, we attempt to distinguish the two cases by detecting one of the particles at detector D.

Now, with our Aristotelian intuition, we might think again, like in the previous case, of two possible cases: the two particles scattering at angle θ or at angle $\pi - \theta$. (We are using radian units from 0 to 2π, instead of angle grades from 0° to 360°.) We might then be tempted to bring detector D ever closer to the small interaction region called the '*collision vertex*' and observe which particle took path i and which particle took path j.

Unfortunately, Heisenberg's uncertainty principle prevents us from doing so! We are not allowed to know what is going on inside this collision vertex, as this is a tiny interaction region where quantum uncertainty reigns. And because we don't know what happens inside this interaction space, we can't tell which of the two particles, A or B, hits detector D. However, it is **not** just like shaking a couple of billiard balls in a black box into which we can't look. As we have clarified while discussing Heisenberg's uncertainty, quantum tunneling and the superposition principles, here we are not allowed to think of the two particles in this region as having exact position and momentum at all, not even in principle.

Here, the wavy nature of the particles, which must be thought of as two wave-packets, are no longer negligible (or, if you prefer the algebraic standpoint, the commutation relations for the position and momentum operators can no longer be ignored). The trajectories inside this region cannot be definite particle paths as our naive intuition would like to consider them. There is no such thing as a point-like particle following a precise trajectory i or trajectory j.

What instead becomes important at these scales are the diffraction and interference phenomena between wave packets. In QM, a scattering process is a diffraction of probability waves. Therefore, we must resort instead to a description with the wavefunction. In fact, physicists say that the wavefunctions overlap and interfere at the collision vertex. It is this that makes particles indistinguishable in QM, not the idea that the observation as such perturbs the system. They are not distinguishable, not even in principle. It is a form of indistinguishability that, again, we must carefully avoid confusing with any classical form of indistinguishability. Particles are not distinguishable because of our ignorance—say, because we couldn't look at the collision vertex with sufficient accuracy. Here, they are indistinguishable because, despite our human reductionist mindset would like to believe otherwise, we must understand the two particles as two wave packets merging and becoming one, and only one whole system without distinctive sub-parts in the interaction region, and which is described by one single wavefunction.

Yet, Fig. 110 represents a still somewhat naive understanding of what is really going on. In Fig. 108 to Fig. 110, we considered only a special case scattering angle θ. If we consider all the possible angles occurring in a scattering process, we must build a wavefunction that also has an angular dependence on θ—that is, at the collision vertex, the two particles must diffract each other in all directions, like a plane wave diffracts at an object's boundaries.

In fact, to simplify this scattering process, consider one of the two interacting particles as a point-like center of wave diffraction (the dot in Fig. 111 left) at rest fixed in the laboratory reference system, and consider the other as an incoming wave packet. If we consider the wave packet of the incident particle to be large relative to this point-like diffraction center, it can be represented as an incident plane wave. Then, once the plane wave hits the localized target region, the resulting wavefunction involves a superposition of an incident plane wave and a scattered spherical wave (Fig. 111 right).

Fig. 111 Scattering as a diffraction of a plane wave by a point-like object and subsequent emission of spherical outgoing wavefronts.

From the perspective of the other particle, the same can be said. It also will 'see' an incoming front and diffract it into a superposition of an incoming plane wave and an outgoing spherical probability wave. The resultant wave is a more or less complicated outgoing spherical probability wave expanding from the interaction center towards the outside world until a similar scattering and interaction process occurs again with another particle.

This is still a somewhat simplified description of the real scattering process because, in reality, both particles should be represented as incoming and outgoing wavefronts, as Ernst Rutherford first introduced to examine the atomic nucleus bombarding it with radioactive particles. (An even more relativistic correct calculation involves a complicated higher order Feynman path integral formulation and diagrams of QFT, which we will mention briefly again later.) However, this should give you at least a qualitative understanding of what QP interactions and scatterings of particles are about.

The probability of finding the scattered particle is no longer given by well-defined and deterministic paths of tiny particles or fields (as in Fig. 108 to Fig. 110) but is determined by a spherical outwardly expanding probability wave.

The situation in its temporal evolution and with the detector induced quantum state reduction is reproduced in Fig. 112. It is still a two-dimensional view. For realistic calculations in three dimensions, we have scattering in all directions not only around a circular region but on a spherical surface. This implies another angular dependence, ϕ, a *solid angle distribution*. For the sake of simplicity, we depict only the two-dimensional section.

The different figures, a, b, c and d, represent different temporal snapshots of the scattering event. In Fig. 112 (a), we have the two incident wave packets of each particle A and B that, before any mutual interaction, are still distinguishable in space. What makes them distinguishable are simply their different spatial coordinates, not any inherent property, as they are two electrons with exactly the same properties.

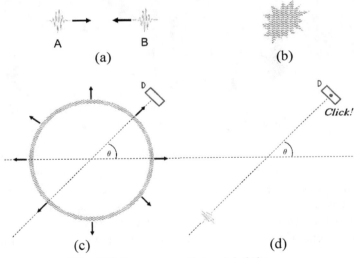

Fig. 112 *Quantum indistinguishability.*

Fig. 112 (b) represents the short time interval during which the two particles interact. Here, the two wavefunctions overlap and the individual spatial coordinates of the individual particles are no longer distinguishable. The two probability waves diffract and interfere with each other in complicated processes. The scattering process is represented graphically simply by a little 'bubble' in space which hides the details of the interaction.

(In QFT, this is also represented by the exchange of virtual particles, as we have discussed in Fig. 103; more about that will be illustrated later.)

After the collision, what emerges are still not two nicely separated wave packets, but a spherically shaped shell wavefunction whose amplitude is modulated by a function that has an angular dependence in spherical coordinates as $f(\theta, \phi)$, also called the 'scattering amplitude'. This spherical probability wave (proportional to the squared modulus of the scattering amplitude) spreads into outer space according to the particles' speed. As long as there is no detection or any other interaction with other particles, this spherical wave proceeds undisturbed and no logical inference is allowed about the particles as distinct entities. This is because, at this stage of the process, there is still nothing like two point-like particles, nor even such a thing as two wave packets—only a spherical probability wave expanding outwards in space.

Only when this spherical envelope reaches the detector and when the detector clicks (Fig. 112 (d)) will the wavefunction collapse back again, by a spacial localization, to a nicely defined spot inside the detector (say, on a pixel of a CCD-camera), which tells us that there, what we call a 'particle' must have interacted with it. Or, as we shall see later and to express it in more rigorous quantum parlance, ,decoherence' takes place which causes the projection of the state vector onto one of its eigenvectors and respective eigenvalues. At the same time, the other particle which emerged along the other path must be visualized again with a wave packet traveling in the opposite direction.

At this point, we no longer have any means of establishing which of the two particles made the detector click. This is for the simple reason that what emerges from the collision vertex is not a system that can be reduced to two separated particles in the first place, but is only one single and unique whole wavefunction spreading out and waiting to be detected. Only when the detector clicks will this 'whole' suffer state reduction and reveal itself again as what we call 'two particles': the one showing up at the detector itself and the other possibly even light years away, on the opposite path, where we could eventually place another detector.

This is something which hints at an inseparable holistic interpretation of reality, as everything in physical reality can be considered a large number of particles continuously interacting with each other. In addition, those particles that interacted here and now form, for a while, an entire unique system which has no parts as long as the next interaction occurs and where this whole again collapses instantly into its particle nature, then again forms another indivisible system with the new particles they encountered on their path. This newly formed wavefunction temporally evolves, spreading out until it finds other particles with which to interact and thereby collapsing again, and so on.

Another typical place to consider this could be the undistinguishable electrons inside the atoms. We already saw that modern atomic theory no longer conceives of electrons as moving along precise and well-defined orbits, but they must be modeled with orbitals. We should not consider orbitals—and, therefore, atoms—as objects made up like a puzzle structure, as this would again imply that the constituent puzzle pieces can in principle be distinguishable from each other. We must instead conceive of atoms as a single, complicated wavefunction, a probability wave that represents a whole, a single unit, which manifests itself in the form of distinguishable different parts only at the instant of interaction or observation but not before that.

To go somewhat further, let us proceed with a bit more formal analysis. We have seen how particles can be in a spin-up-down state superposition. In the same manner, there can also be a superposition of scattering states.

States in superposition for a single particle are described by a state vector as a sum of the possible eigenstates, as, for instance, in Eq. 21. On the other hand, the vector state of a system of many particles is described by the product of the possible eigenstates of each particle. In our present case, consider particle A which follows path i as being in state $|i\rangle_A$ and the other particle B following path j as being in state $|j\rangle_B$ (see Fig. 110 left). Then we can represent the state vector of the system as the product of the two states as: $|i\rangle_A |j\rangle_B$. The same goes for the other situation in which particle A follows path j and particle B goes along path i. Then the overall state vector is : $|j\rangle_A |i\rangle_B$ (see Fig. 110 right). However, if this quantum algebra were correct, this would still allow for distinguishability, as even for identical particles, the formalism distinguishes between the two cases. In this sort of quantum algebra, a classical indistinguishability is still encoded.

A real quantum indistinguishability must also take into account the superposition principle. Hence, we must allow also for our intuitive understanding of the quite disturbing eventuality that each particle travels both paths at the same time! That is, the state vector of the whole system must describe both particles traveling both paths at the same time. This is given by the sum (or subtraction) of the two possible states above, that is:

$$|\Psi\rangle = \frac{|i\rangle_A |j\rangle_B \pm |j\rangle_A |i\rangle_B}{\sqrt{2}}. \qquad \text{Eq. 34}$$

The square root of two is simply a normalization factor that stands for the probability coefficient (see Eq. 11) in which the modulus square represents the probability of getting the first or second case ($|\frac{1}{\sqrt{2}}|^2 = \frac{1}{2} = 0.5 = 50\%$ chance of getting the first or second case each). The $\pm$ sign corresponds to two different particles: a positive symmetric wavefunction for bosons (as we know, particles like photons with integer spin s = 0, 1, 2, ...) and a negative anti-symmetric wavefunction for fermions (particles like electrons, protons, neutrons, etc. with half-odd integer spin s = $\frac{1}{2}, \frac{3}{2}, \frac{5}{2}$ ...). We will discuss the meaning of the negative sign later in detail; please accept this little ambiguity for now.

So, we have taken two steps. First, we represented the system of two particles taking one OR the other path as the product of their respective states. Then we added (or subtracted) the two states as a superposition of two possible scattering states—that is, the particles taking both one AND the other path at the same time. By doing so, any notion of distinguishability fades away. Not only is it impossible to distinguish the two electrons after the scattering process, but the question no longer makes sense. If both electrons travelled along both paths at the same time, the single electron measured at detector D is simply another 'entity' that emerges from the collision vertex, which is the spherical wavefunction. What detector D measures is a particle, which, of course, preserves exactly the same properties of the incident ones but can no longer be considered either A or B (or, maybe A and B are the very same electron?). Again, QP reminds us that any logical retro-duction implying counterfactual definiteness is a bad habit of the human mind. This is the real quantum indistinguishability. It is just another form of quantum superposition.

One might ask, at this point: Why should we take the above-mentioned second step, that of a superposition of scattering states, seriously? After all, it offends our common sense to consider that particles travel distinct paths at the same time. Can we not simply skip that passage and write the state vector as $|\Psi\rangle = |i\rangle_A |j\rangle_B$ or $|\Psi\rangle = |j\rangle_A |i\rangle_B$ and be happy with that? The proof that Nature instead insists on taking superposition seriously is an experimental proof. You will be able to predict the correct scattering rate, as well as the probabilities of how the electrons are deflected over all the possible angles θ and that are empirically observed with laboratory experiments, if and only if the calculations are done with the wavefunction of Eq. 34. Any other description fails to describe reality.

In conclusion, generally speaking, what we must bear in mind is that indistinguishability does not emerge because of our ignorance or the imprecision of the measurement devices. Nature insists on the indistinguishable quantum version at all costs. There is no perturbation effect or measurement imprecision preventing us from distinguishing between them. Rather, quantum indistinguishability emerges due to the superposition of possibilities. The simple fact is that there is nothing to distinguish, and there are no differentiated parts to differentiate in the first place.

This is the base, the conceptual foundation, of a closely related phenomenon that has puzzled—and that continues to puzzle—so many bright minds and upon which we will now build in the coming section: the phenomenon of quantum entanglement.

2. Quantum entanglement basics

Let us connect the concept of quantum indistinguishability to another important concept, that of *quantum entanglement*. If you got the message about why and how quantum indistinguishability is very different from our ordinary conception of indistinguishability, it should no longer be difficult to grasp the quantum entanglement concept.

We will now see how this sort of 'quantum wholeness' not only is a phenomenon that stands behind indistinguishable particles but in QM is a more general and fundamental fact, also between very different particles, like the entanglement between an electron and a positron, or a proton and a neutron, or as is usually done experimentally nowadays, simply between two photons but with different polarizations.

So, what is quantum entanglement? The term was coined by Schrödinger from the German word 'Verschränkung'. Physically, quantum entanglement can be realized when particles such as electrons, protons, photons or even larger molecules interact with each other quantum mechanically at a microscopic scale and then separate. After separation, as long as they do not interact with another particle, atom or molecule in the environment, they remain entangled, just as we saw in the case of quantum scattering and the spherical probability wave. As long as one of the electrons does not interact with the detector, the compound system is not described by the wavefunction of two particles or even not by two wave packets but by a unique undifferentiated whole expanding into space, or simply a state which one must describe as 'undefined'.

A first possible example of producing entangled photons is that obtained by using radioactive materials. Consider a radioactive source of decaying atoms. We know that this is a typical quantum process (recall what we have learned on quantum tunneling). A typical radioactive decay is the β^+ decay ('beta-plus decay'), whereby a proton in an atomic nucleus decays into a neutron and emits a positron (the anti-electron with positive charge and opposite spin), together with the emission of other particles, that is, an electron-neutrino, and eventually gamma rays (see Fig. 113 left).

The neutron remains inside the atomic nucleus and converts the atom into another element (one with a proton less, with atomic number decreased by 1). The positron is instead ejected out of the atomic nucleus and, unless the environment is not a perfect vacuum, will soon find on its way another electron and annihilate by emitting two photons in the opposite directions and with opposite circular polarizations (with opposite helicity, that is, spins), as shown in Fig. 113 right.

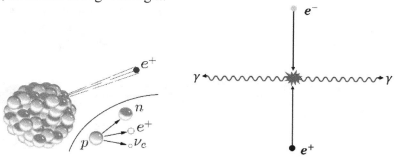

Fig. 113 Left: The radioactive β⁺ decay.
Right: Entangled gamma photons from the electron-positron annihilation.

This is just an example with photons, as photons are usually the most manageable object in the laboratory. Many other possible sources of entangled particles are possible. So, let us keep general and think of particles (photons or material particles), which are emitted by a radioactive decay or by a source from which they had a common origin. They are emitted in exactly the opposite directions because the originating decaying single particle is at rest in the laboratory and, therefore, because of the universal law of the conservation of momentum and energy, they must fly in directions which cancel their momenta. For the same reason, the two particles must also have opposite spins, as the original particle from which they originated had opposite intrinsic angular momentum which sums up to zero momentum. Therefore, the two spins must cancel each other as well; that is, if one particle shows up with spin-up, we can be sure that the other will show up with spin-down. This means that we have only to look at one particle's spin to know that the other particle will have the opposite spin. Moreover, think of them moving apart to a great distance (say, giving them sufficient time to eventually separate light years away) before making a measurement on it. During this time, we must also imagine them having, along their path, no further interaction with anything else in the environment. The particles must maintain their state of coherence (that is, formally this implies that the relative phase between the particles' kets in the wavefunction remains constant in space and time).

Now, as we are accustomed to doing in this book, let us first take the naive classical understanding and then shift to the correct quantum description of what is going on.

Naively, before any measurement, we imagine the system to be in one OR the other state, that is, that of the following two possible cases.

As first case I (see Fig. 114), we have a particle with spin-up travelling towards the left side. Let's label it direction A. On the opposite side, in direction B, we have the particle with spin-down. Then, in the exact same fashion as we did in the section on quantum indistinguishability, we can represent this system using a product state vector $|+\rangle_A |-\rangle_B$.

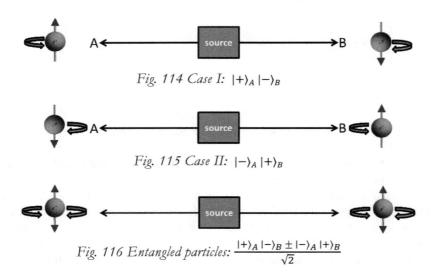

Fig. 114 Case I: $|+\rangle_A |-\rangle_B$

Fig. 115 Case II: $|-\rangle_A |+\rangle_B$

Fig. 116 Entangled particles: $\dfrac{|+\rangle_A |-\rangle_B \pm |-\rangle_A |+\rangle_B}{\sqrt{2}}$

The other possible case II is that in which we have the particle on the left with spin-down, and the particle flying in direction B with spin-up (see Fig. 115). Therefore, the state vector must be written as $|-\rangle_A |+\rangle_B$. We can have only these two cases. We can't get both particles with spin-up or down; they must be '*anti-correlated*', as is customary to say. But, we get to know this only after the measurement. Before the measurement, we don't know which of the two possible outcomes will be measured.

However, again, this is a naive classical understanding of things. The two states, those of case I or case II, are not what QM considers the state of the system during flight time because the particles would, in principle, still be distinguishable.

In QM, indistinguishability is much more radical than the classical one and extends into all quantum phenomena, not just scattering processes. The same fits in the present situation. We aren't allowed to distinguish between which particle goes left or right; therefore, we aren't allowed to distinguish between which has spin-up or spin-down as we would like to imagine naively with case I or case II. This must be expressed again in the algebraic expressions of QM. When we have still not measured the system, during flight time, QM forces us to take the superposition principle seriously. That is, it wants us to consider the superposition of both possible cases I and II, that is, the situation depicted in Fig. 116.

Therefore, the full quantum state $|\Psi\rangle$ of the system must be described by a single state vector, of particles with opposite spins and zero total angular momentum, also called a '*singlet state*', which is a sum of the two possible eigen-states of the two particles:

$$|\Psi\rangle = \frac{|+\rangle_A |-\rangle_B \pm |-\rangle_A |+\rangle_B}{\sqrt{2}} \qquad Eq.\ 35$$

with the $\pm$ sign again representing the case of bosons or fermions, respectively, as we shall see later.

So, according to QM, during the flight we have neither case I nor case II, but both at the same time. We again face the logical AND case. However, when the measurement is performed, the state vector is projected onto one of the two eigenvectors, that is, $|+\rangle_A |-\rangle_B$ (the particle on the left appears with spin-up and the particle on the right appears with spin-down) or $|-\rangle_A |+\rangle_B$ (the particle on the left appears with spin-down and the particle on the right appears with spin-up), each case with 50% pure chance. Note that in Fig. 116 the labels A and B were omitted to highlight indistinguishability.

Once a measurement of the spin occurs on one particle, state vector reduction takes place. If we measure a spin-up state for the particle on the left side, we already know that, even without measuring it, the other particle must possess a spin-down state and vice versa. That is, the act of measurement collapses the entangled system of particles either in the state of case I or case II.

The point is, according to QM, this state reduction, or wavefunction collapse, happens instantly. The state vector projection is an instantaneous process. The 'click' of a detector, which determines the state of one particle, automatically and instantly also determines the opposite state of the other particle. Again, it is not that we 'discover' the state of a particle, which we didn't know because of our ignorance; the two particles are in a superposition state as long as no measurement takes place. Only once the detector clicks will the measurement 'define' the state of one particle. Also, only then will the other particle's opposite state become defined too.

A common fallacy is to believe that the fact that the outcome is unpredictable and that we always have a 50% pure chance of measuring one or the other state implies that we can believe that the particles intrinsically possess one or the other state when the source has emitted them. Resorting to a classical analogy, this is like the situation of two tossed coins that are separated light years away from each other and in which we look at them only after space travel. According to this view, based on a counterfactual definiteness reasoning, some very complicated process is going on at the source emission of the particle whereby we must describe their states with statistical probabilities, but they nevertheless supposedly possess a definite anti-correlated state also during the flight time but that we will know only by measuring it. This, however, would contradict quantum indistinguishability, which we know correctly predicts scattering processes as measured in the laboratory only once we accept that when two equal quantum objects interact with each other, they can no longer be considered a separate system made up of two or more particles. They 'melt' into one single quantum entity, a sort of unique undifferentiated whole that can not be represented by any sort of reductionism. Moreover, it would imply that QM must be a theory with hidden variables. We will see, however, that the idea of a quantum theory containing hidden variables is possible only in the context of a theory which must accept non-local instantaneous interactions.

We must conclude that, in the case of the entangled particles, their respective state is decided only at the time of measurement, not before that instant. This justifies also the name 'singlet', as we are dealing with something which is supposed to comprise two particles and yet behaves as a single entity. Entanglement given by Eq. 35 can be thought of as the realization of a physical state that sums up the principle of superposition, the principle of indistinguishability and the uncertainty principle.

At this point, you may already have sensed the big problem with all that. If the instantaneous state reduction that defines the state of one particle must instantly determine that of the other particle as well, even if they are light years apart, this seems to imply an FTL interaction!

Keeping entangled particles far apart is experimentally a very difficult thing to do, as any particle will soon react with another particle in the environment, causing decoherence. Conceptually, however, it is a perfectly legitimate physical possibility. Let us take, for instance, the case of a particle measured on Earth and another, entangled one which has reached another stellar system, say, on the Alpha Centauri star, 4.3 light years away. This instantaneous state reduction (or 'collapse' of the wavefunction, if you prefer to call it that) implies that a measurement on Earth immediately determines the state of the particle on Alpha Centauri. The question, then, is: How can a particle on Alpha Centauri immediately know the state of the measured particle on Earth taking its opposite spin state? A light signal from Earth to Alpha Centauri needs 4.3 years to transmit this information.

Obviously, there is a huge problem with all this because Einstein's theory of relativity strictly forbids FTL interactions. That the speed of light is a physical limit which cannot be overcome is a fact that several experiments have confirmed. If that were not the case, the entire theory of relativity itself would be false too, though it has been tested and shown to be correct over and over again throughout over a century of experimental evidence.

You can understand the kind of feeling this must have instilled in Einstein, the father of the theory of relativity. Einstein called that strange behavior of entangled particles a *"spooky action at a distance"*, and he obviously could not accept that at all. There seems to be a conflict here between QM and relativity.

Even after his debate with Bohr, Einstein could not accept this state of affairs, at least not in this form. He was very puzzled by the idea that, in the quantum reality, things could be entangled, seemingly through a FTL correlation and with totally random outcomes as the theory describes. He simply could not believe all that to be true. Somewhere, somehow, there is something that is escaping our understanding. He continued pursuing the idea that all this must be only a provisional description of reality, which, however, cannot be complete and must lead to contradictions. Einstein was looking for something which shows us that we must return to a classical view, or at least to an understanding which saves, we might say, the two coins tossing representation of reality—that kind of reality in which objects have inherent properties independently of the act of measurement.

At some point, it seemed that Einstein, together with a few of his colleagues, found a powerful argument, again, in the form of a thought experiment which seemed to be a final proof that Heisenberg's uncertainty principle indirectly implies an FTL interaction among quantum entangled particles, and that this could not be compatible with relativity. This is the famous paper describing what nowadays is known as the 'EPR paradox', which will be the topic of the next section.

3. The EPR paradox (original version)

It was the year 1935. Einstein's debate with Bohr saw the latter emerge as the winner. Yet Einstein did not give up. In a famous paper entitled *"Can quantum-mechanical description of physical reality be considered complete?"* [13], Albert Einstein, Boris Podolsky and Nathan Rosen (EPR), proposed an interesting thought experiment which was supposed to show why QM cannot be complete after all, as, so their argument went, it inherently contains a statement which denies physical reality (if, by 'physical reality', we mean any physical quantity that can, in principle, be determined with certainty).

To show what they felt was inconsistent with the present formulations of QM, EPR chose position-momentum complementarity.

We know that if we precisely measure the position of a particle, we will lose information about its momentum and vice versa—if we measure its momentum, we can no longer say something precise about its position, not because of an interaction between the measurement device and the particle, but because of a fundamental, inherent indefiniteness of these quantities intrinsically, in themselves. I hope you now grasp this important distinction.

Fig. 117 Albert Einstein   Boris Podolsky   Nathan Rosen
(1879-1955)   (1896-1966)   (1909-1995)

So, what was the EPR argument, which, so they believed, shows that there is a possibility of overcoming this limitation? If you understand the real meaning of the complementarity between position and momentum expressed by their commutation relation, it is easy to understand what they had in mind by using the following measuring scheme applied to a system of entangled particles.

Suppose we have a source, which we assume to be at rest relative to our laboratory, that produces, by a nuclear decay process, two entangled particles which fly apart into two different and opposite directions (see Fig. 118).

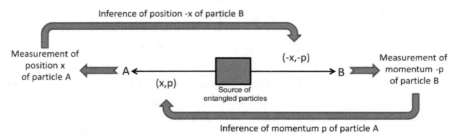

Fig. 118 The EPR thought experiment.

Then, according to our standard intuitive understanding of Newtonian dynamics (which, as everyone knows, implies the conservation of energy and momenta), we should think of each particle as having the same but contrary momentum, that is, p for particle A and -p for particle B. Now measure the position x of one particle (say, particle A, for example) by inserting a tiny pinhole in its path, or whatever, but without making any attempt to gain information about its momentum p because, according to QM, we are not allowed to know the position and the momentum at the same time. Otherwise, once we measure the momentum, we will destroy the information over its position again. So, we perform only one position measurement on particle A and get the measurement result x.

With this information, we should have also established the position of particle B, at least according to our common sense, as it was ejected from the source at the same time and with the same momentum in the opposite direction. Therefore, at the time of the measurement, the position of particle B must have been -x. We can deduce this without any measurement, simply by an inference which relies only on the assumption that the conservation of energy and momenta holds.

The interesting point is that, because so far no measurement of particle B has been made, QM allows us to make a second measurement: the measurement of the momentum of the second particle B. (Measuring momenta is somewhat more complicated from a practical point of view, but not impossible in principle.) As long we make the second measurement on the second particle B, QM has no objections. Now we have the momentum -p of particle B and, again for the conservation laws, we deduce from that, without any measurement, that the momentum of particle A must have been p. Therefore, by this thought experiment we can, at least in principle, retrieve the position x and momentum p of particle A and also the position -x and momentum -p of particle B without having to measure both position and momentum on the same particle—and this, apparently having deceived Nature, which imposes the commutation relation on these observables stating the impossibility of knowing them both, not even in principle.

Note that because the position of A and the momentum of B have been inferred mentally, by deduction, not from a measurement, the implication is that, if Nature nevertheless maintains Heisenberg's principle as valid, the uncertainty cannot arise due to the perturbation or disturbance of the system for the simple reason that no measurement (of p on A and -x on B) has occurred in the first place. This also means that if the particles' positions and momenta are really definite, as EPR argues, then the wavefunction and the entire algebraic construct describing a quantum system lack the complete information, as it allows for only one or the other dynamic variable to be determined, according to Heisenberg's uncertainty principle and the commutation relations.

By this simple reasoning, EPR concluded as follows: *"If, without any way disturbing a system, we can predict with certainty (i.e., with probability equal to unity) the value of a physical quantity, then there exists an element of physical reality corresponding to this physical quantity." - "While we have thus shown that the wave function does not provide a complete description of the physical reality, we left open the question of whether or not such a description exists. We believe, however, that such a theory is possible."* [13]

We can sum this up by stating the following.

"If the condition for the reality of a physical quantity is the possibility of predicting it with certainty without disturbing the system" then

1) "the description of reality given by the wavefunction in quantum mechanics is not complete"

or

2) "these two quantities (x and p) cannot have simultaneous reality".

This is known as the '*EPR paradox*'. The paradox is not in the logic or the math, but in what this logic suggests in contrast to our physical macroscopic everyday experience. Nature seems to opt for the second possibility with QM complete, contrary to any form of realism to which we are accustomed.

Moreover, EPR argued that if QM is a complete theory, this must imply a causality violation in the relativistic sense. There can be causes and effects which arise apparently due to an interaction faster than the speed of light. This is because, if one measures the position of particle A and, by doing so, discovers that this instantly causes an uncertainty over the momentum of particle B, even if it is light years away, this looks like an interaction at a distance that is superluminal, in the same manner we already saw in the previous section with quantum entanglement. Because relativity was already an experimentally tested and accepted theory, in EPR's view, QM had to be modified. It could, at best, be considered only a provisional theory and not a complete one. If QM imposes the non-commutability of operators as a fundamental principle, a thought experiment like that which EPR proposed should not be allowed to circumvent it.

What EPR were looking for was a local theory—that is, a theory that does not need apparently instant superluminal correlations over long distances and that does not deal with intrinsically undetermined states. They were looking for a theory which would reveal how QM in its present form must be incomplete and which ultimately explained facts in terms of predetermined and well-defined states, as in Newtonian mechanics or relativity, and which rested on the assumption that hidden variables exist that rule the phenomena even if they have yet to be found. In this thought experiment, the predetermination they were seeking was some mechanism, some still-unknown process, which poses the two particles into an opposite spin state at the time of their creation, already at the source, and that could dispense with the requirement of any annoying FTL correlation.

EPR were apparently able to show that the principle of complementarity expressed by commutation relations seems to be a formal mathematical abstraction, but cannot describe how things are really, objectively, 'out there'. Physical objects must ultimately have intrinsically definite properties, even if for whatever reason we can't measure them directly. That is, EPR tried hard to reintroduce counterfactual definiteness into the physical picture. In the EPR thought experiment, one deduces properties, such as the position of particle B and the momentum of particle A, not measuring it but by a retro-ductive reasoning that assumes, more or less implicitly, that particles inherently have a position and momentum, even if we are not able to know them. QM, however, seems to shed serious doubts on that. The question at this point is: Were EPR really right? Is QM an incomplete theory after all?

4. The EPR paradox (modern version)

The EPR paradox in its original form, with the position-momentum measurement, is quite difficult to make. For a long time, it remained a thought experiment that was impossible to realize. The advent of modern technologies, lasers and quantum electro-optics devices was necessary to make it feasible. Most of the modern experiments testing QM rely on the measurement of the commutation relation of the spin observables. (In fact, throughout the modern literature, you will usually find the EPR experiment explained straight away with the particles' intrinsic angular momentum.) For this reason, it is worthwhile to reformulate the EPR paradox into its modern version.

It was originally David Bohm who translated the EPR paradox into its spin analogue. He proposed using electron-positron pairs that are created by a decay process (more precisely, by a neutral π-meson, a sub-atomic particle that can be produced in an accelerator collision process and that quickly decays in an electron-positron pair) and that have a total of zero angular momentum, that is, zero total spin. However, lasers are technically an easier and cheaper option and modern experiments use photons, that is, particles with integer spin. To produce the entangled photons, physicists use laser beams that excite the atoms of some particular substance (usually Beta-Barium Borate (BBO) crystals) in such a way that it re-emits the absorbed energy in the form of entangled photons and with special polarizations (see later the special section dedicated to photon entanglement) or helicity, that is, spin.

Whatever kind of technology one uses to produce entangled particles, the important point to keep in mind is that in the case of an entangled system, you will never know which of the two particles will show up with the 'up' and 'down' spin (or, for photons, the left- and right-handed polarization).

To visualize EPR's thought experiment with the spin operators, suppose the two entangled particles with opposite spins and zero total angular momentum (that is, in a singlet state) have been produced by some source and fly apart in opposite directions towards two observers. Let us call them Alice and Bob, like in Fig. 119.

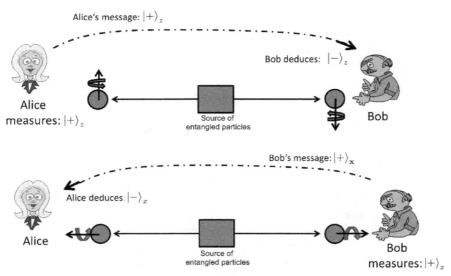

Fig. 119 The EPR experiment with particle spins.

If the two particles are entangled and if Alice will measure (say, along an axis z) a spin-up, $|+\rangle_z$, she can send this information to Bob, who, without any measurement on his particle, will deduce that it must be in the opposite spin state, namely, the $|-\rangle_z$ state. Now he can instead measure along a different axis, for example, the x-axis. Let's say he gets a $|+\rangle_x$ spin state.

Then he returns the favor and sends a message about his particle state to Alice. Therefore, she concludes that the state of her particle along the x-axis should be $|-\rangle_x$, again, without having performed such a measurement on her particle.

At this point, both might believe that they have the full set of information about the two observables for their own particles, the spins along the x and z axes, but having measured only one of them on each particle, while deducing the other indirectly after being informed through some standard communication channel. Alice's particle must be in a $|+\rangle_z|-\rangle_x$ state, while Bob's particle must be in a $|-\rangle_z|+\rangle_x$ state, and neither of them had to measure the particles along both axes.

However, this is not allowed according to standard QM, which imposes the commutation relations on the spin operators and which tells us that it is strictly forbidden to obtain both spin states at once because: $[s_z, s_x] \neq 0$.

So, do the two particles really possess objectively these spin properties along the two axes at the same time? We already saw with the Stern-Gerlach experiment that this is highly doubtful, to say the least. In any event, the problem with all this here is that, if we want to answer this question, we will have to measure Alice's particle along the x-axis, while Bob will also have to make a second measurement along the z-axis. We know that the laws of QM strictly forbid this, and we know how it would end if they did so: It would again render indeterminate s_z for Alice and s_x for Bob. The only thing that is sure is that a strong correlation exists between the two particles.

So far, we have seen the EPR paradox in its original version with the position-impulse determination of a couple of particles and in its more modern version, which resorts to the spin of particles. Both versions are substantially equivalent in the sense that they seem to imply that the commutation relations of QM could, in principle, be violated. If so, this would vindicate Einstein's strong conviction that QM must somehow be an incomplete theory and that a more refined and better theory must exist, possibly based on hidden variables, which could explain things better without paradoxes.

However, when it appeared, this very interesting paper of EPR did not attract attention at all. Even though one of its authors (Einstein) was a globally known, high-ranking physicist, almost nobody cited the article, and it remained completely ignored for thirty years. This was quite characteristic of the spirit of the time, during which almost all physicists embraced the Copenhagen interpretation and stopped thinking about these more foundational and philosophical aspects of QM. Einstein was considered an old man who'd had his moments in the sun but who was no longer able to produce new and original contributions.

Yet history has shown that EPR were probably too far ahead of their time. The fact that the technology required to perform EPR's experiment was still not mature was not the only reason why they were ignored. It was an evolutionary necessity, the culture of the time, and minds were not ready to focus on these subjects.

This changed in the late 1960s with Bell's theorem. But, before we analyze EPR's legacy in the light of Bell's intellectual achievement as the logical continuation of EPR's spirit of investigations into the foundations of QM, let us first focus on the issue of FTL transmissions.

5. Faster than light transmissions?

Let us briefly address a topic which you might have heard about frequently: that of a supposed possibility of taking advantage of quantum entanglement, or whatever other quantum effects, to transmit information at speeds FTL, or eventually also instantly throughout the Universe via a wavefunction collapse of anti-correlated particles. This is something to which many like to resort in order to explain possible telepathic or paranormal phenomena. To see how this is unfortunately not the case, not even in principle, let us connect again to the modern experimental setup of the kind we used for the EPR paradox, then further follow the considerations and consequences to which it leads.

In its spin version, with Alice and Bob making measurements on the spin of the two entangled particles flying apart, we saw that, once the spin of one of the two particles is measured, the spin of the other particle will immediately be determined.

We also saw why in QM (at least in its standard formulation that is considered a complete theory and without hidden variables), we are not allowed to think that this correlation is predetermined; that is, it does not arise among the particles at the instant of creation in the source, but must be understood as being determined at the time of measurement, not before. We cannot use counterfactual reasoning, whereby we imagine them as being one particle in a spin-up state and the other particle in a spin-down state during flight while we are not looking at it. We must consider the two particles as a one-and-unique whole in a superposition state of the two possibilities, and that only when the measurement takes place will one particle assume the spin-up and the other the spin-down state, just by pure 50% chance. Therefore, while the overall momentum in its singlet state is zero, the anti-correlation of the two particles comes into existence at the time of measurement.

Here, a very appealing idea naturally comes to mind. We might use this superluminal action at a distance to send messages throughout the entire Universe at speeds FTL, maybe even at infinite speed everywhere, contrary to all that relativity has told us so far. That could eventually also be a great method of communicating with extra-terrestrial civilizations, couldn't it?

So, the question really is: Could Bob and Alice use this method to instantly transmit messages to each other throughout the Universe? Moreover, the two spin states of elementary particles lend themselves quite well to serving as a binary encoding system, like that in computers. Nature did us the favor of allowing for particles having only discrete two valued spin states. For example, one might imagine digital data in the form of ones and zeros associating to the spin-up the 1 digit and to the spin-down the zero symbol. Could we, in principle, use this binary spin encoding to transmit instantly, say, with Alice on Earth and Bob on a planet around a star in the Andromeda Galaxy and which is 2.2 million light years away?

According to QM, at least as we know it in its present version, the clear answer is: no! Not even in principle.

The point we must always bear in mind is that neither Alice nor Bob will ever be able to predict or force the outcome of the measurement they make on their particles. They are not allowed to intervene in the random quantum processes which entangle the particles at the source and they cannot predict or influence the measurement outcome when they receive and measure their particles in their respective laboratories. They are simply passive observers of a stream of random digits and have no means of projecting their particles into a desired set of eigenstates. Who decides where the spin-up or spin-down will appear is neither Alice's nor Bob's business. It is the exclusive competence of Nature upon which we have no authority.

In fact, to transmit any sort of information, would require, say Alice, forcing with some non-random regularity the spin outcomes onto her entangled particle. Otherwise, she cannot encode any message in a binary code.

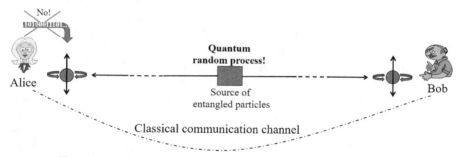

Fig. 120 Alice can't predetermine the outcome of her measurement!
To know the particles' complete state a classical communication channel is needed!

Yet this is forbidden! She has no means by which to intervene and decide into which eigenstate her particle will be projected at the time of her measurement. She can only make measurements and find out which spin sates show up for each particle, but she cannot predetermine these spins of the particle. What they both see is a random stream of events produced by the radioactive or whatever kind of entangling source, but they will never,

ever be able to predetermine the next particles' state. This prevents, from the outset, any possibility of encoding data. According to QM, there is always a 50% chance of obtaining the 'up' or 'down' spin, and there is nothing you can do about it, other than passively listening to white noise. If they can't determine which spin value comes into being at the instant of measurement, they have no means by which to transmit information.

Therefore, even if there is, indeed, a correlation between two particles coming into being instantly over long distances, it will nevertheless be impossible to use it to send information. For this reason, there isn't any real violation of the causal process in the relativistic sense. It is a non-local causality but in line with relativity if we restrict our understanding of relativistic causality to something which is about only slower-than-light-speed information transmission. At any rate, the quantum FTL correlation must not be seen as a transmission of particles or a force or a field that spreads out from Earth to the Andromeda Galaxy instantly. All the known particles, forces and fields that physics deals with are known to always propagate with speeds equal or less than the speed of light, in accordance with relativity. It is the wavefunction that collapses instantly and determines anti-correlation. Yet the wavefunction is not an EM field or a wave made of particles or a force. So, what is it? Well, nobody knows for sure. It is just a probability wave. We can't say more than that and therefore we should not extrapolate to particles or fields traveling FTL.

Moreover, recall how in this experimental setup, which is the analogue of the EPR experiment extended to particle's spins, both Alice and Bob have to ask each other about their measurement results if they want to know the complete state of their particles, because QM does not allow them to perform further measurements on their own particles. Bob has to ask Alice what she obtained along the z-axis and Alice has to ask Bob what outcome he got along the x-axis. Only after receiving this information can they infer the complete state of their particles. To do so, they must use a communication channel—say, a contact through a radio telescope or whatever kind of classical communication channel that cannot transmit signals faster than the speed of light. They must both wait a couple million years for their respective 'cosmic telephone call' to effectively note an anti-correlation of the spins. Before that, Alice will not know anything about the x-spin value of Bob, while Bob will know nothing about the z-spin of Alice's particle.

So, finally, while it might well be true that something in the quantum world looks like determining correlated events among particles instantly, apparently by an FTL action, this cannot authorize us to believe that it can be used for FTL communication. Correlation does not automatically imply transmission of information. In this sense, there is no contradiction between QM and relativity. Relativity is safe insofar that it imposes the no FTL information transmission. We must distinguish between a superluminal causation and a superluminal information exchange. The former looks to be commonplace in QM, whereas any eventuality of the latter must be excluded.

The non-local character of QM is frequently invoked to explain psychic phenomena such as telepathy, etc. However, you cannot resort to QM to justify FTL telepathy or mind reading, at least not QM as it is known in its present formulation. Despite so many contrary claims in the popular media, nothing in QM allows for that. If this would one day turn out to be the case, another mechanism in our brains must be at work beyond simply an EPR correlation. In fact, some scientists are doing serious research to determine whether, in biological processes, quantum effects may be relevant. These scientists wonder whether the brain is subjected to quantum phenomena. (We will discuss this topic in the second volume.) However, you should be aware that, at this stage, these are still very speculative hypotheses which must be confirmed. Therefore, if you want any serious physicist to take you seriously, please realize that any statement connecting QM to paranormal phenomena must be taken with skepticism or at least great care and discrimination.

6. Schrödinger's cat paradox, quantum decoherence and the measurement problem

We are now ready to understand the famous (perhaps too famous) *Schrödinger's cat paradox*. We saw that superposition implies, in the strange world of QM, that objects can be in more than one quantum state simultaneously. For example, in several places at the same time, or spin clockwise and anti-clockwise at the same time. If you assume that you understand all this, then, as American physicist and Nobel laureate Richard Feynman used to say, you probably don't understand QM. This totally paradoxical Universe confused and eluded any

Fig. 121 Erwin Schrödinger (1887-1961)

reasonable explanation by the best minds of the 20th century. Until his last days, Einstein could not accept this state of affairs. This prompted other great scientists, such as the Austrian physicist Erwin Schrödinger, to propose a slightly eccentric thought experiment.

Schrödinger wondered about the weirdness of the quantum superposition principle. The question to which he was seeking an answer was: Why do we not observe objects in the macroscopic world, in our every day experience, in superposition states? It seems that these strange and counterintuitive properties, such as superposition, emerge only in the tiny microscopic realm. And yet there is no reason, nor any known physical law, that forbids the same effects from taking place on larger scales, for objects as big as a human body or ... a cat. To better visualize the problem, Schrödinger used the cat experiment to illustrate why there is no reason to believe that superposition should not hold equally well for cats as it does for elementary particles. He reasoned as follows.

Imagine a cat in a closed box, having no interaction with the outside world. In the same box is a radioactive substance, which at any instant can emit a particle due to atomic decay. We have learned that radioactive decays are also strictly probabilistic phenomena; quantum tunneling determines the decay of an atom's nucleus. Because this quantum tunneling is a purely quantum phenomenon, you can't ever determine when an element decays before you observe it, just as you can't say that a particle has a specific position or spin state before you measure its position or intrinsic angular momentum. So does the element which finds itself in a superposition of decayed and non-decayed state.

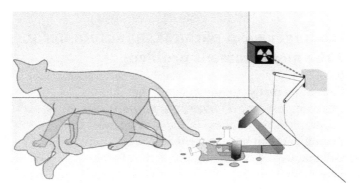

Fig. 122 The Schrödinger's cat thought experiment.

However, from the statistical point of view, we can predict how much average time is needed until half the atoms do decay – the so-called 'radioactive half-life time'. This tells us the probability that decays will take place and how many decays, on average, will take place after some time. Therefore, we will never, ever be able to predict when exactly the single decay event will take place. Schrödinger imagined that the box also contains a Geiger counter, an electronic device that 'clicks' when it detects a particle emission. When it measures a decay, it releases a hammer that breaks an ampoule containing poison and then kills the cat. But, this will occur only if that one atom of the radioactive substance decays. Because this is a perfectly quantum probabilistic event, as long as we do not open the box (the wavefunction still has not collapsed), according to the rules of QM, the cat, the radioactive material, the hammer and the ampoule containing poison must be in a superposition state. This means that the ampoule is broken and intact at the same time, the hammer is released and not released at the same time, and therefore the cat must be dead and alive at the same time! Only when we open the box to check on the health of the animal does the wavefunction collapse, which means we make the observation, the act of measurement, and 'project' the cat in either the dead or alive state. However, before the instant of observation, QP seems to suggest to us that the cat, indeed, exists in a sort of ambiguous reality in which it is alive and dead at the same time.

In fact, as in the case of the single elementary particle, such as an electron, we know that it can be in a coherent state as long it does not undergo a measurement which projects it on one of its definite eigenstates. The same applies to 'quantum cats'. As long as a measurement does not take place, the cat finds itself in the superposition dead AND alive at once. This is formalized in QM with the state vector as the sum of the representing ket states over the normalization coefficient whose modulus square gives a 50% chance for one or the other case happening. Once the measurement has taken place, the system – here, the cat in its superposition state – is projected onto one or the other axes representing a definite eigenstate, dead or alive.

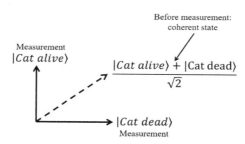

Fig. 123 Schrödinger's cat superposition state.

With this thought experiment, Schrödinger could show that it is not at all clear why we do not observe superposition states in the macro-world, as this strange possibility certainly exists in the microscopic quantum world. At least in principle, this experimental set up shows that it is possible to have quantum superposition states for beings and objects the size of a cat.

Did Schrödinger find a loophole in this concept? If QM is true, we should be able, at least in principle, to perform such an experiment and put cats in the dead-alive superposition state. However, the fact is, in our everyday life experience, nobody has ever observed this. So, where is the mistake? Are we perhaps working with some false assumption?

Schrödinger's idea is only a thought experiment, which can't really be performed, not so much because of technical limitations but because one must first open the box to see in which state the cat is. However, this is an act of measurement which will cause the collapse (or state reduction). Therefore, we will never be able to determine how the cat felt before the box was opened. Even if the cat survives, it can't tell us because cats don't speak. Guess what? Why not use a human instead of a cat and, after the experiment, ask him or her how it feels to be in a dead and alive superposition?

Fortunately, we do not have to go that far to understand why there is no real paradox. However, we must first elucidate one of the most frequent misconceptions about QP, one which is frequently stirred up by many popular media (as well as by some physicists who are eager to sell their books): the supposed 'role of the observer in QP'.

This little phrase has caused many misunderstandings. What is meant by an 'observer' in QP? What do we really mean by the word 'measurement' in QP and how should we distinguish it from the word 'observer'? After all, an observer measures, doesn't it? It is necessary to be cautious about not playing with words which might lead (accidentally or intentionally) to several misunderstandings.

A 'measurement' in physics is, after all, nothing but a physical information readout, which usually (but not always, as we shall see later) is carried out via an interaction with the system to be measured. Is it really necessary that a measurement be linked to the observation of a conscious being, like a scientist, a human being looking upon something? The point is (and this will turn out to be of paramount importance when people speak about the consciousness of the observer influencing the outcome of a quantum experiment), in QP the notion of 'observation' and 'measurement' must be intended as being extended to ANY possible interaction of one system onto another.

No physicist in a laboratory has ever directly observed a quantum phenomenon at a microscopic scale. Every measurement is not an observation they perform with their sense of sight or touch, but a readout performed by electronic devices like photodiodes, CCD-cameras with laser beams, etc. The wavefunction collapse (or state reduction) is caused by a measurement that these devices make and that has no contact whatsoever with the observer's mind or consciousness. Should we consider the photodiode a 'conscious observer'? Can a thermometer measuring the temperature in a room be regarded as an 'observer' that observes temperature willingly and consciously? That is, if we would also like to think of an act of measurement performed by a mechanical apparatus as an 'observation', it must be clear that the mind or consciousness of the scientist using these devices to perform the experiment has nothing to do with what causes the outcome of the measurement. Therefore, this unfortunately widespread myth of the supposed 'role of the human mind or consciousness in QP' is only a misconception that has no scientific or logical ground. We will take up this issue in the coming chapters and see how QM is a contextual theory, which means the quantum behavior of a system is influenced by the setup of the experimental context (in some sense, we have already seen this with the Stern Gerlach experiment, haven't we?), but there is no role of the consciousness, the mind or the emotional state of a human observer.

'Observing' in the quantum mechanical sense also means the interaction of one particle onto another. There is no logical reason to believe that the interaction between two particles or between a molecule and an electron is physically and conceptually different from the interaction between two physical objects, one of which is the object to be observed and the other of which is the measurement device. Measurement is always about the interaction (or information transfer) between two material objects. In QP, by an act of 'observation', we must not intend a conscious being or only a macroscopic device, but must also include the interaction of a particle with another particle. Every single interaction of the gazillion particles in our environment is, according to QP, a 'measurement'.

So, does this imply that a single particle could collapse the wavefunction of a quantum system? Well, things are a bit trickier than that. Several particles can also be entangled, and in quantum superposition while they are interacting with other particles, but this does not necessarily immediately lead to what we imagine as the wavefunction collapse. In fact, to have the Schrödinger's cat experiment succeed, not only must the tiny atom be in a superposition state $|\Psi\rangle = \frac{|decayed\rangle + |not\ decayed\rangle}{\sqrt{2}}$ but this superposition must also hold, at least for a short time, for all the atoms and molecules of which the Geiger counter, the hammer, the poison, the ampoule, the walls of the box, the air molecules in the box and especially all the cells of the body of the cat are made, with them all being quantum entangled in a unique, undifferentiated and coherent whole. 'Quantum coherent' means that the phase relation between the wavefunctions describing all the constituents does not change (recall the characterization of quantum coherence and de-coherence we applied in I.3 to Eq. 1). Because nobody has ever observed such a weird thing, there is a conceptual disconnect between what QM suggests microscopically and what we experience macroscopically. This disconnect is called the *'quantum measurement problem'*.

That this state of affairs cannot be taken so lightly in QM is illustrated through an application of Dirac's notation to a system that transitions from a superposition to a definite state – that is, to one of its eigenstates. Consider a system in a generic superposition, such as:

$$|\Psi\rangle = \frac{|a\rangle + |b\rangle}{\sqrt{2}} \, ,$$

with $|a\rangle$ and $|b\rangle$ being, for example, spin-up/down or cat alive/dead, or two mutually exclusive, that is, orthogonal physical states. If we want to calculate the probability that the quantum system will change from a state $|\Psi\rangle$ to one of the two possible states, say $|a\rangle$, we must apply Eq. 18 and write (see how to calculate the modulus squared in the appendix with *Eq. 46*):

$$
\begin{aligned}
|\langle\Psi|a\rangle|^2 &= \left|\left(\frac{\langle a| + \langle b|}{\sqrt{2}}\right)|a\rangle\right|^2 = \left|\frac{\langle a|a\rangle + \langle b|a\rangle}{\sqrt{2}}\right|^2 \\
&= \frac{1}{2}(\langle a|a\rangle + \langle b|a\rangle)^* \cdot (\langle a|a\rangle + \langle b|a\rangle) \\
&= \frac{1}{2}|\langle a|a\rangle|^2 + \frac{1}{2}|\langle b|a\rangle|^2 + \frac{1}{2}\langle a|a\rangle^*\langle b|a\rangle + \frac{1}{2}\langle b|a\rangle^*\langle a|a\rangle \\
&= \frac{1}{2}|\langle a|a\rangle|^2 + IntTerm(\langle b|a\rangle) = \frac{1}{2}|\langle a|a\rangle|^2 \, . \quad Eq.\ 36
\end{aligned}
$$

Well, this all looks very cumbersome and strange. However, if we take a closer look, we find that it reveals a deep physical meaning. First, the last three terms of the second line have been shortened for convenience as an 'interference term', IntTerm($\langle b|a\rangle$), because it contains the combined inner products of state $|a\rangle$ and $|b\rangle$. It can be called an 'interference term' because it is the same type of interference term we encountered in the double slit experiment in Eq. 1. (After all, it is the same calculation, but with interfering quantum states $|a\rangle$ and $|b\rangle$ instead of interfering light beams $|\Psi_1\rangle$ and $|\Psi_2\rangle$.) However, this has been set to zero in the last passage for the simple reason that $\langle a|b\rangle = \langle b|a\rangle = 0$, because the inner product of orthogonal state vectors is zero (recall Eq. 18 or how we encountered orthogonality for spin states in Eq. 25) and what remains is just what we already know, namely, that there is a 50% chance of getting state $|a\rangle$. This is fortunate! Otherwise, those remaining terms wouldn't make much sense. If the inner products $\langle b|a\rangle$ were not equal to zero, it would mean that there is still a non-zero probability that we could measure the system still being in two states at the same time, against anything our worldly experience suggests.

However, this is true only for a system in isolation. The fact is, these conditions are rarely met. Any system that is not in a perfect vacuum will quickly somehow interact with its surrounding environment. Moreover, if we perform a measurement with some macroscopic device, we must also take into account all the interactions with it. This means all the quantum states of the system to be measured will become entangled with the virtually infinite number of particles of the environment plus the measurement apparatus. Formally, this implies that the eigenbasis of the quantum system, in our case $|a\rangle$ and $|b\rangle$, by its interaction with the external macroscopic world, will become entangled with the environment and the apparatus as a whole, that is, with its eigenbasis $|\phi_i\rangle$, where the index i runs over the virtually infinite number of eigenstates. The general environment and apparatus quantum state could be represented as $|\phi\rangle_{EA} = |\phi_1\rangle|\phi_2\rangle|\phi_3\rangle$ ... Then the quantum state of the system entangled with the macroscopic world could be written as:

$$|\Psi\rangle_{EA} = |\Psi\rangle \cdot |\phi\rangle_{EA} = \frac{|a\rangle|\phi\rangle_{EA} + |b\rangle|\phi\rangle_{EA}}{\sqrt{2}}.$$

Repeating, then, the calculation above all over again with this last term (which is quite cumbersome because it entails a multiplication of brackets running on the index i, which we omit), one obtains an analogous result as *Eq. 36* but extended to the environment and apparatus as:

$$\left|_{EA}\langle\Psi|a\rangle_{EA}\right|^2 = \frac{1}{2}\left|_{EA}\langle a|a\rangle_{EA}\right|^2 + \text{IntTerm}\left(_{EA}\langle b|a\rangle_{EA}\right),$$

where we use the shorthand $|a\rangle_{EA} = |a\rangle|\phi\rangle_{EA}$ and $|b\rangle_{EA} = |b\rangle|\phi\rangle_{EA}$ and with the interference term that, this time, we cannot sweep under the carpet so easily by setting it to a zero value because $_{EA}\langle b|a\rangle_{EA}$ is the inner product of interacting eigenbases, that is, non-orthogonal bases. What this seems to suggest to us is that, at least for a short time, the quantum system must be in its superposition state, which is truly entangled with the external world. There is effectively a quantum interference between the object and its surroundings. This is another way of seeing quantum coherence extended to quantum systems that we imagine are made of many particles. In other words, we should truly observe the cat alive and dead at the same time.

However, the question is: How short is that 'short time' of coherence? One can show that this loss of coherence, the so-called '*quantum decoherence*', occurs on time scales that are inversely proportional to the number of particles involved and decreases exponentially with time. The number of molecules which make up the air, the cat's body and the surrounding objects is of the order of Avogadro's number, a number which tells us how many constituents a mole of a substance is made of. (For instance, a mole of water is about 18.3 grams.) Avogadro's number is $6,022 \cdot 10^{23}$! And if you think about how these constituents interact with each other continuously due only to the thermal motion of the atoms and molecules, the so-called '*Brownian motion*', it becomes clear that the decoherence time must be vanishingly small only because of the interaction with the thermal excitations in the environment. This is called *'thermal decoherence'*.

So, according to this understanding of what is going on, macroscopic objects could also, in principle, be in superposition or in an entangled state. However, the time interval of this 'state of being' is so ultra-short, it can be considered, for all practical purposes, as instantaneous. Quantum decoherence sets in extremely quickly and we won't notice anything strange that might suggest anything other than our good old classical world.

In principle, to avoid decoherence, we must be able to perfectly maintain the coherence of all the particles of the environment and every measurement device by eliminating the thermal agitation. This, in turn, would practically mean that any measurement device could not function in the first place and that the poor cat must be frozen to an absolute zero temperature (which would at least solve the dead-or-alive dilemma from the outset…).

So, when we consider huge macroscopic systems, which are not cooled down to absolute zero temperature and are not placed into very special physical conditions, we will never, ever observe quantum coherence – that is, anything in quantum entangled superposition. It is not that the laws of QP no longer hold; they hold at all levels. However, for larger objects, all the quantum effects disappear almost instantly. Nowadays, exceptions to this can nevertheless be realized with the so-called '*Bose-Einstein condensates*' (which we will be described in the selected advanced topics book) by complicated procedures placing a gas of helium in a container at fractions of degrees Kelvin avoiding its contact with the chamber walls. However, there is no way to conceive of a living cat's body as being in the same physical conditions to experience the dead-alive condition.

So, finally, the so-called 'Schrödinger's cat paradox' is not a paradox at all. It can be explained with quantum decoherence taking place in very short times because of the interaction between the huge number of particles of which a macroscopic system is made and the particles of the rest of the environment. This is, at present, the most accepted explanation. Schrödinger's contribution was to point out that there is no reason to believe that, at least in principle, quantum superposition is a scale-dependent universal feature of physical reality.

However, the story still does not end here. One also hears professional physicists, not particularly informed in the foundations of QP, say that the decoherence process solves the measurement problem. This is not completely correct. It provides an explanation of why Schrödinger's cat paradox is not a paradox, but it does not entirely solve the measurement problem.

Decoherence explains why we perceive objects as being classical, together with '*Bohr's correspondence principle*'. According to this principle, CP is a limiting case of QM. That is, QM reproduces CP in the statistical limit of large quantum numbers. (Recall how the increase in energy quantum numbers behaves in the limit as large orbits or large energies, as we saw in Bohr's atomic model, or how quantum numbers determine the orbitals at the end of section II.9.) However, decoherence solves the measurement problem only partially, as it still leaves open an 'anomaly' that is intrinsic in the formalism of QM.

Note that decoherence is, strictly speaking, not the usual wavefunction 'collapse'. Rather it is, so to speak, a quick 'interference suppression'. As fast as it might be, it should be described by an evolution in time described with the evolution operator, the unitary operator we discussed in section II.9 (see Eq. 16 and Eq. 17). If so, the same unitary operator U must describe the time evolution from the state $|\Psi\rangle_{EA}$ to $|a\rangle_{EA}$ as also to $|b\rangle_{EA}$ because these are two different possible outcomes of the same initial state. In formal language, this implies that if $|a\rangle_{EA} = U|\Psi\rangle_{EA}$ and $|b\rangle_{EA} = U|\Psi\rangle_{EA}$ then (because of Eq. 17 and Eq. 14):

$$_{EA}\langle a|b\rangle_{EA} = {}_{EA}\langle\Psi|U^*U|\Psi\rangle_{EA} = {}_{EA}\langle\Psi|1|\Psi\rangle_{EA} = {}_{EA}\langle\Psi|\Psi\rangle_{EA} = 1.$$

However, this means that, indeed, we have the certainty to observe the two eigenstates in superposition also at the macroscopic environmental scale of the measurement apparatus (or the cat)! This is contrary to the previous statement that, due to decoherence, the interference term is rapidly suppressed, that is, $_{EA}\langle a|b\rangle_{EA} = 0$.

This contradiction is somehow intrinsic to the mathematical description of QM and emerges when we resort to decoherence only as the solution to the measurement problem. It isn't. The issue is still debated among physicists and perhaps even more among philosophers because it does not really prevent the carrying out of calculations and precise predictions on the microscopic scale with elementary particles, as physicists are busy with. Some of the interpretations of QM, such as the many world interpretation, the De Broglie-Bohm pilot wave or the objective collapse interpretation (see the selected advanced topics edition), could, in principle, resolve this contradiction. However, these interpretations, while they reproduce the predictions of QP and are empirically indistinguishable from the orthodox theory (which is why they are called 'interpretations') also suggest mutually exclusive ontologies. Nobody knows which must be preferred and whether any of it at all has something to do with reality. Maybe this shows us that Einstein was not so wrong. QM might not be so complete after all. However, whatever resolution the measurement problem might find inside an alternative complete description of reality, it certainly will not be that for which Einstein had hoped.

VI. Digging deeper into the quantum realm

1. Bosons, Fermions and Pauli's exclusion principle or: why is matter 'hard'?

Asking why solid objects, such as a granite stone, convey a feeling of impenetrable compactness or asking why we can't walk through a wall, might seem at first to be silly questions. However, what is this 'hardness' that makes solid objects impenetrable and difficult to modify? What kind of forces act at a microscopic level to prevent our bodies from falling towards the Earth's center?

These seem to be futile questions only because we are so accustomed to some properties of the material world in our daily life, and therefore our so-called 'common sense' has never given it a thought. And yet, if you think about it, you will find that these questions do not have straightforward and obvious answers.

The first answer might be that all material objects we perceive with our natural senses are made of atoms and molecules that are kept tightly together by atomic and molecular bonds in more or less regular crystal lattices that resist attempts to break them apart. For example, Fig. 124 shows the lattice of frozen water molecules, commonly called 'ice', which everyone knows can form a quite rigid physical object.

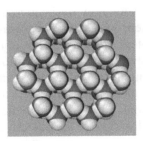

Fig. 124 Naive view of an ice lattice.

These kinds of pictures, which are supposed to represent the different atoms or molecules, can be found almost everywhere in chemistry textbooks. However, they are somewhat misleading from the quantum mechanical perspective: There are no such solid tiny 'balls'. The pictures tend to suggest the existence of rigid bodies because the mind has the habit of thinking about it as such; it is simply a naive way to perceive and think of microscopic objects with the same mental process we have used since childhood. We have already seen that we must think in terms of atomic orbitals, not 'billiard-ball-like' particles.

Furthermore, we should not forget that atoms are almost empty objects. As we already found out in discussing the Bohr atom model, almost all the volume of an atom is empty space. The size of a proton is confined to a space of the order of femtometers (1 fm $=10^{-15}m$) but the electron cloud is 100,000 times larger than that. To understand the relative sizes, imagine a proton as big as a marble of 1 cm in diameter. Then the electron cloud diameter of a hydrogen atom is 1 km in comparison. Because over 99.94% of the mass is concentrated in the nucleus, it is pretty clear that atoms are far from being compact and solid objects.

What we must always keep in mind when dealing with atomic bond or molecules or interacting particles is that the rigidity we experience in everyday life is due to intra-molecular forces and not to direct contact between atoms and molecules. There is no 'contact' among constituents in the classical sense of which we think.

Fig. 125 reminds us of the difference between a naive understanding of the water molecule (left), in which people imagine three hard and solid 'marbles' sticking together, and the more realistic representation of QP with an orbital model, that is, a probability wave through which to find the electrons bound to the oxygen and two hydrogen nuclei. There are no such things with a definite and sharp boundary (and even less those ridiculous light shadings), but only a diffuse, smoothly varying probability-electron-cloud.

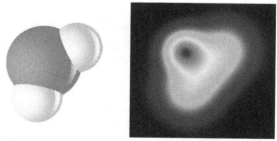

Fig. 125 The naive H_2O molecule model and its probability-wave orbital representation.

So, what makes these objects collectively feel solid and hard when they stick together into a larger crystal lattice? The same question arises when we think of collisions, such as that between two billiard balls. When two objects collide, is there also a 'direct collision' or 'direct contact' at the microscopic scale between the particles which make up the atoms and molecules of the colliding objects itself? What, after all, do the words 'direct' and 'contact' and 'collision' mean in QP?

Let us think about what really happens microscopically when two elementary particles 'bounce' onto each other. To a certain extent, we already answered this question when we dealt with the scattering process in chapter V.1 in elucidating the quantum indistinguishability principle. We saw, in that context, that it is all about mutual interactions between two force centers, that is, between their force fields. However, in this case, we are speaking of two macroscopic objects whose quantum effects are not expected to play a role. Do the molecules of the billiard balls, their atoms and the particles that build them up really collide, 'touching' each other? We might imagine again the fundamental constituents at the interface where the billiard balls collide and interact, again, as a contact between two hard, solid objects that undergo an elastic or inelastic collision. However, this implies that we are once again thinking about them as non-point-like objects, which have a finite size and a certain extension, and possibly themselves as being made of other particles and sub-constituents. So, this reasoning leads us to an infinite regression. We have only shifted the problem, not solved it.

Then let us conceive of real point-like elementary particles, just mathematical points with zero volume and size. A point has no structure. Can there, then, be any collision in the classical sense, as we are accustomed to conceiving of it intuitively? You might easily realize that there can't be anything like that. That is because if these points are infinitely small, an infinitely precise shot is necessary to make them collide. Or, to put it in physics parlance, in the case of a composite object made of several point-like particles with zero extension and no volume, the mean free path which defines the average distance between two successive collisions of a couple of point-particles would be infinite. This is quite intuitive if you think that the ratio between their mutual distance and their zero-size gives an infinite number. Thus, we should see objects go through each other, and we should be capable of walking through walls, etc.

This is a much less weird and unrealistic state of affairs than you might think. Paradoxically, we have a very nice example of this in the domain of astrophysics and cosmology. Think, for instance, of galaxy encounters. Our Sun is an average-size star (about 110 times the diameter of the Earth), but it is nevertheless extremely small when one compares it to the size of the Milky Way, the galaxy it hosts. With their telescopes, astronomers have taken nice pictures of lots of interacting galaxy encounters. The stars are so small compared to the entire galactic structure that when two galaxies 'collide' with each other, in reality, the probability that one star of one galaxy will physically collide with a star of the other galaxy is almost zero. Both galaxies will merge and penetrate each other. Two stars almost never come directly in contact; however, when they eventually come sufficiently close, they swing around each other, getting pulled by their respective gravitational forces. In a galaxy encounter, the two galaxies will indeed be deformed by their gravitational fields, but will then either continue along their path or form a larger structure, with several stars scattered throughout space. But no, or almost no, star will have suffered a direct physical impact with another star. In fact, because there is no real collision despite the 'galactic crash', one speaks of a *'collision-free interaction'*.

Fig. 126 A collision free interaction between Galaxies NGC 2207 and IC2163.

This macroscopic cosmological example might have suggested to you the answer we were looking for in a microscopic domain of elementary particles. What really makes two particles deviate from each other is never a real collision of two solid objects which have precise and definite boundaries, like billiard balls. After all, we know that this kind of reasoning is forbidden in QM by the uncertainty principle. Everything in the quantum world should be considered 'fuzzy'. The reason why particles act onto each other is that they interact through some sort of force field that surrounds them. In the previous case, it was the gravitational field of a star. The greater the mass, the greater the field strength. (The Earth's gravitational field might be taken as a more familiar example.) However, in the particle domain, the gravitational force is so weak, it can be neglected. What is much more important in QM are the electric and nuclear forces. Take, for instance, the electric field between two opposite electric charges. The positive- and the negative-charged particles will attract each other, and in physics, it is customary to depict these fields with lines. For two opposite charges, the particles feel an attractive force and the field lines are directed towards each other, as shown in Fig. 127 left.

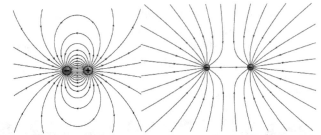

Fig. 127 Electric field lines between two opposite and two equal charged particles.

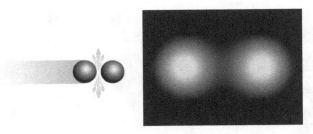

Fig. 128 Naïve understanding of particle collision and vs. quantum field interaction without the particle concept.

If the two particles have the same electric charge (say, two electrons with negative charge), they repel each other and one depicts the field lines going away from each other, as shown in Fig. 127 right. However, these graphical representations do too much to satisfy our naive understanding of the world, as we still imagine two solid particles with definite size and boundaries. And, of course, there are no 'lines' in the real physical world, though they help our imagination represent force fields. These models are still based on an everyday experience which sees 'marbles' collide and touch each other as in Fig. 128 left, following precise trajectories. In the quantum domain, it is much more appropriate to imagine particle collision as zero-dimensional mathematical points in space acting as centers of a force field like in Fig. 128 right. The particles are surrounded by a smoothly varying force field. Near their respective centers, the force becomes so strong and the repulsion effect is so large that, in fact, we could consider their mutual interaction as a 'collision' or a direct 'contact' between two apparently solid objects, leading to a scattering process. However, these 'solid objects' never exist; they are only a sort of mental reconstruction of our mind, a mere concept with which we are accustomed because we are living in a macroscopic reality. Pushing together two electrons will strengthen the repulsive forces, mimicking a collision between two solid extended objects. However, this is only an illusion of our senses and a false understanding of what is really going on. In other words, 'hardness' is an emergent property, not an inherent quality of particles.

If there is an interaction only between two force centers, and never a direct 'contact' or collision' as we might think of, that explains what happens when we touch material objects or hit an obstruction. We can't walk through a wall, or fall towards the Earth's center, because at a microscopic level, the electrons that make up all the atoms and molecules of our bodies, as with any material object we may encounter, all have the same negative electric charge and therefore repel each other through their electric force fields. The sum of all the gazillions of tiny electric repulsions between the electrons produces the macroscopic net reaction force that we perceive in our daily lives when touching objects or hitting our heads against a wall. Ultimately, it is a collective electric phenomenon which allows us to take things in our hands. However, in truth, there is no 'touching' at all; there is only a play of mutually interacting force fields.

But what about neutrons, then – that is, the particles which have no electric charge and reside in the atom's nucleus? Can we conceive of them as point-particles? And can we think that because there is no electric repulsion, they might be pushed together so that they come so near to each other, we might consider them as being practically in the same place at the same time? It turns out that this is not the case at all.

Take as an example the helium atom, as in Fig. 129. This is surrounded by an electron cloud, as we discussed for the orbital model of atoms. Going deeper into its structure, we find two positively charged protons plus two neutrons, apparently neutral. I say 'apparently' because neutrons are not really neutral. They are electrically neutral only at large distances, in a far field approximation, say, at about orders of a magnitude of the atomic diameter (at about one Ångström, that is about a tenth of a millionth of a millimeter).

However, their electric neutrality comes from an internal structure composed of even smaller constituents with opposite charge: the famous quarks (as are protons). There are different types of quarks; a neutron is made up of one 'up' quark and two 'down' quarks (these are just labels and have no meaning whatsoever), while a proton has two 'up' quarks and one 'down' quark. However, quarks have an electric charge: the down -1/3 and the up +2/3 the charge of the electron. This implies that while a neutron is globally neutral in the far field, in the near field, it is made of an electric charge distribution.

Moreover, neutrons (like protons) have a force field that surrounds them, one which not only is the electromagnetic force but which also comes from strong nuclear forces responsible for atomic nuclei remaining bound despite the protons having the same positive charge. Therefore, two neutrons, when they approach, will 'feel' each other's electric and strong nuclear fields.

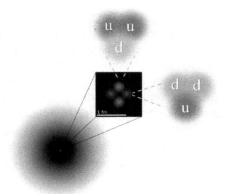

1 Å = 100,000 fm = 10^{-10} m

Fig. 129 Structure of the Helium atom and its nucleus.

The process describing this combined interaction is quite complex and can be stated only by modern QFT. However, what results from these calculations is that almost all the mass of protons as neutrons is due to their strong force binding energy. Recall that according to Einstein's energy-mass equivalence ($E = mc^2$), the energy content of an object contributes to its mass. For example, if we extend or compress a spring, it accumulates potential energy and gains mass. In principle, we could note that the spring weights more in a state of tension. However, of course, the effect is so minuscule that even with the most sensitive and precise measurement device, nothing can be observed. This is not the case for the strong forces in nuclei. This force deserves its name entirely. It binds the quarks inside the atomic nuclei (as in all hadrons, that is, all the particles made of quarks) so strongly that the mass of the proton and neutron is due almost entirely to its binding energy content. In other words, not only is the atom almost void with all the mass concentrated in the nucleus, but also that mass is again 99% due to the foam of energy of the nuclear strong force field. Only a tiny 1%, if any, could be considered 'mass' in the classical sense.

This should give us a better reason to keep in mind that, also, here nothing can be conceived of as particles with a definite size and boundary 'colliding' and 'touching' other particles in the classical sense we like to imagine. There are only complicated fields of force centers.

Instead, electrons are considered to be real elementary particles, that is, they have never been shown to be composite objects and can be thought of as simply points with no size and structure. Electrons feel the electric force but are not subjected to the strong nuclear forces, though they are subjected to the weak nuclear force which can convert them into other particles, such as a neutrino.

And what about that tiny particle which is the neutrino? A neutrino does not couple with the strong nuclear forces either and has no electric charge. That is the reason why it is able to go through entire planets and stars without being affected. It is not because of its small size; instead, it is the lack of interaction. For instance, neutrinos are created in the center of the Sun due to thermonuclear reactions which occur at temperatures of millions of degrees. A proton-proton reaction forms a heavy hydrogen nucleus, deuterium, plus an anti-electron (the positron) and a neutrino. It then travels straight through the entire star without even 'noticing' its existence (contrary to photons, which need millions of years to get through, as we have learned with the Compton effect). Once it reaches the Earth, in most cases it simply flies through it without interaction as well. The main reason we know of its existence and are able to detect it is that it nevertheless 'feels' the weak nuclear forces, that kind of nuclear force responsible for the decay of the atomic nuclei. On extremely rare occasions, neutrinos, when flying near an atomic nucleus, can cause atomic decays, which allows us to detect them indirectly. Otherwise, the neutrino would be a mere theoretical hypothesis, a ghostly particle we could never be sure actually existed.

So, our conclusion must be that elementary particles mimic a solidity because they interact with each other through force fields. No marbles or billiards in sight.

Let us continue with this line of reasoning and answer another question. Do particles exist that could be in the same place at the same time? Intuitively, it might look like an impossibility, and yet you know these particles very well from your everyday experience: photons. Light particles are electric and magnetic fields oscillating, but have no net electric charge, have no mass, and are not subject to nuclear forces. We have no problem with conceiving light rays overlapping each other. In fact, it is normal practice in physics to represent electromagnetic fields which add up in the same point of space at the same time linearly, like every other wave does – for example, like water waves or sound waves. (Well, water waves do not always add up linearly; special waves called 'solitons' exist that exhibit another behavior, but let us not digress). For example, when we observed the fringe interference pattern on a screen, we inserted the screen just on the intersection of two light rays, as in Fig. 130. However, without the screen, the light rays – that is, the photon particles – are perfectly able to intersect and overlap without perturbing each other.

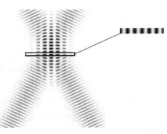

Fig. 130 Photons can be in the same place at the same time: superposition of two waves and the interference pattern at an inserted screen.

It is perhaps somewhat less intuitive to imagine two material particles occupying the same space than it is for their wave counterpart. However, as you know by now, there is a wave-particle duality and therefore no reason to think that things should be otherwise in the particle picture. It is legitimate to conceive of photons as particles that can be in the same place at the same time.

As we have discussed above, the impenetrability, the 'hardness', the impossibility of two material particles being in the same place at the same time is due to the fact that, when they are near each other, they interact through a force field. However, photons are the electromagnetic force carriers that make up the field itself and are not confined by any force field.

As we will see in the following sections, the interactions in QFT are often represented by special types of graphs: the '*Feynman diagrams*', which show how the interaction between particles is mediated by the exchange of force-carrying particles (for instance, photons). The typical example of the scattering process between two electrons is represented by the exchange of a photon which carries the momentum from one particle to another. Fig. 131 shows two electrons coming from the left. Once they are sufficiently near to each other, they exchange a photon which carries an impulse and causes them to deflect. This is a quite different understanding of how things work as compared to what we said for Fig. 128 right, isn't it?

In fact, we already made it clear that photons are one example of a more general class of particles called 'bosons' (from the name of the Indian physicist Satyendra Nath Bose, 1894-1974). We saw how they are distinct from another class of particles, namely, the fermions. Bosons have all integer sized intrinsic angular momentum, that is, spin $S = 0, \pm 1\hbar, \pm 2\hbar, \pm 3\hbar, ...$

Other bosons exist, because the strong and weak nuclear forces also exist, and as gravitational forces. The bosons of the strong nuclear forces are called '*gluons*' because their function is to 'glue' and bind together the quarks, protons and neutrons in the nucleus.

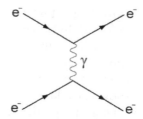

Fig. 131 Electron-electron scattering process.

The mediators of the weak nuclear forces are the W^+, W^- and Z^0 bosons with spin $\pm\hbar$, while the boson for gravity is the '*graviton*', which has never been observed directly as a particle but whose existence is not difficult to imagine because the existence of gravity is a quite obvious fact, and it is expected to have spin $\pm 2\hbar$. There is also the famous Higgs boson, which is a particle that, in the frame of modern QFTs, in the standard model of particle physics, explains why particles have a mass, and has 0 spin, a so-called '*scalar particle*' in contrast to '*vector particles*', which have a non-zero spin.

The other class of particles, those which interact with each other through the bosonic force fields but which are distinct from bosons, are the fermions (from the name of the Italian physicist Enrico Fermi, 1901-1954). All the material elementary particles of which we are made – electrons, protons, neutrons and the quarks which build up the latter two, and which in turn build up the atoms, molecules and the 'hard' stuff that makes up the Universe we observe and perceive with our senses – are fermions. You should know by now that these have a half-integer-sized spin $S = \pm\frac{1}{2}\hbar, \pm\frac{3}{2}\hbar, \pm\frac{5}{2}\hbar, ...$ Even neutrinos are fermions. However, keep in mind that the notion of bosons and fermions is usually extended beyond elementary particles. It is customary to speak of bosons, gas of bosons or bosonic matter of any composite object that has a net integer spin and fermions, gas of fermions or fermionic matter otherwise.

Recall also that bosons and fermions differ in another essential aspect. Fermions possess that bizarre property which is exquisitely quantum mechanical: if you rotate them by 360°, they will not return to their original quantum state. To re-obtain a fermion in its initial state, you must rotate it by 720°. A similar strange property of fermions is what is called '*exchange anti-symmetry*'. If a system is made up of two perfectly identical fermions (say, two electrons), which for the principle of quantum indistinguishability must be considered perfectly identical, and then exchange their position (that is, you shift the left electron to the right and the right electron to the left), the obtained quantum mechanical system does not return to an identical state!

What is obtained is a state equivalent to that in which one of the two particles has been subjected to a rotation of 360°. The wavefunction describing the state of the two electrons again acquires that little minus sign we already saw for spinors because of rotation. Meanwhile, a particle exchange, that is, a particle permutation, in a system of two bosons exhibits the same quantum state. So, the Universe we know is made of bosons which exhibit the normal intuitive rotational and exchange properties, and fermions which instead must be rotated twice to return to the original state and exhibit exchange anti-symmetry.

However, conceptual caution is advised here. The two anti-symmetric behaviors, the rotational and the exchange anti-symmetry of fermions, are not directly related. To fully explain this would involve sophisticated mathematical and conceptual considerations of modern relativistic QFT far beyond the scope of the present introduction. However, according to a '*spin-statistics theorem*', one can show how this is a direct consequence of three fundamental aspects of the physical reality in which we live: that the Universe has at least three or more dimensions, that the theory of relativity (in particular relativistic causality) holds, and that the laws of QM are true (quantum commutation rules hold). Then only two types of particles can exist: namely, bosons and fermions, with integer and half-integer spin, respectively. It is also interesting to point out that if we lived in a Universe with only two space dimensions, spins could be fractional or even irrational. It is difficult to imagine what kind of Universe, if any, would emerge from such physical conditions.

This also has deep implications in terms of how we write the state vectors for bosons or fermions. We know that a two-particle system (here, one boson in a spin-up state and the other in the spin-down state) can be represented by the product state vector: $|+\rangle|-\rangle$. Then exchange symmetry – that is, exchanging their spatial ordering – implies that for bosons:

$$|+\rangle|-\rangle = |-\rangle|+\rangle,$$

which means the state vector of a quantum system made of two bosons does not distinguish which of the two bosons is in which spin-state. Exchanging the spatial order of the bosons leaves the state invariant. After all, this aligns with the principle of quantum indistinguishability.

Not so for fermions. In this case, exchange anti-symmetry is represented by a minus sign on the right-hand side of the previous state equation. For fermions, one has:

$$|+\rangle|-\rangle = -|-\rangle|+\rangle = e^{i\pi}|-\rangle|+\rangle,$$

where the last term, once again, helps us avoid confusing the subtraction symbol with the spin eigenstate symbol. It tells us that fermion exchange induces a π-phase shift – that is, a 180° phase shift in the product state. (If that looks mysterious to you, check out the complex numbers mathematical primer in the appendix.)

However, if the exchange between two particles does not lead to the same system, this could, at least in principle, allow us to distinguish the two particles, which the principle of quantum indistinguishability strictly forbids. We already know which trick Nature has devised to re-establish the 'indistinguishability dictatorship': it adds up both possible states by quantum superposition; that is, it asks for the state to be an entangled system. The two-particle system is not in one or the other above-mentioned states. Rather, it is in both states at the same time. The quantum state vector of the composite system must then be:

$$|\Psi\rangle = \frac{|+\rangle|-\rangle - |-\rangle|+\rangle}{\sqrt{2}}, \quad Eq.\ 37$$

with the square root over two being, as usual, the normalization constant which modulus square accounts for the probability of observing the first or second outcome. This is the '*singlet state*' (the only possible state with total spin 0), and can be found throughout physics textbooks, though frequently without an explanation for that seemingly mysterious minus sign. It simply accounts for the exchange anti-symmetry of fermions.

The fact that under spatial permutation of boson and fermion particles the state vector of the first does not change sign, whereas for the latter it does, indicates that the wavefunction of bosons must be symmetric, while for the fermions it must be anti-symmetric. (See also the analogy with the sine and cosine function in Appendix A.Ic.) Fig. 132 illustrates this fact. Imagine two particles inside a potential well, forcefully enclosed and trapped in some limited region of space, which means that they are also mutually interacting and therefore entangled. Then the two bosons or fermions will behave and be described by two different and opposite wavefunctions, independently from the kind of forces acting between each other. The two-bosons quantum state in the potential well will be described by a symmetric wavefunction, which means it will not change if mirrored at the origin axis. (The numeric values of the real part of the wavefunction, $Re[\Psi]$, on the left and right sides of the vertical axis of the graph of Fig. 132 left, are always the same.)

On the other hand, the two-fermion quantum state is given by an anti-symmetric function, that is, it is anti-symmetric with respect to mirroring the function at the origin axis. (The numeric values of the real part of the wavefunction, $Re[\Psi]$, on the left and right sides of the graph of Fig. 132 right, will change sign.)

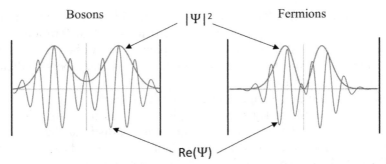

Fig. 132 The symmetric and anti-symmetric wavefunctions of bosons and fermions.

The modulus square of the wavefunctions will then represent a probability density to find one or another boson or fermion in a given region (the $|\Psi|^2$ curve) inside the potential well. (Here we consider the simple one-dimensional case; the horizontal axis represents the position of the particles while the vertical axis gives the probability density to find it in some place.) Though the squared modulus of a symmetric or anti-symmetric function is always symmetric, the two cases reveal a substantial and decisive physical difference. For bosons, the two probability density functions assigned to each particle overlap in the origin; thus, there is a non-zero probability of finding both bosons in an interval in that region. That is, there is a chance that you will find both bosons in the same place at the same time.

Not so for fermions. The two probability-density functions for the two particles never overlap; at the origin, there is zero probability of finding them, that is $|\Psi(0)|^2 = 0$, always. This implies that two or more fermions will never occupy the same place at the same time.

With this, we have found yet another reason why some particles exhibit the 'hardness' property, which is due not only to the action of the microscopically acting fundamental forces (EM forces and the two nuclear forces) but to the wavefunction anti-symmetry which causes them to remain separate from each other.

The natural question at this point is: What happens if we take fermions (say, electrons) and push them together with increasing force? Well, first of all, we must overcome the electrostatic repulsion force between the electrons themselves. This is easier said than done because a tiny amount of matter has an incredibly large number of electrons and would need a gigantic compressive force to squeeze the electrons into an increasingly smaller volume of space. Can we find, in the Universe, some force capable of doing that? Sure. It is the gravitational force of collapsing stars in their lifetimes' end phase. When stars have consumed their nuclear fuel after a long time of stellar evolution, they collapse under their own weight because of gravity. Then, due to a complex astrophysical process, they explode in a supernova and/or leave behind a remnant star, a so-called white dwarf, or a neutron star, or even a black hole depending on the collapsing star's initial mass.

White dwarfs (in Fig. 133, the little white point at the lower left of the brighter star, which is Sirius, the brightest star of our earthly sky) are stars which, thanks to their intense gravitational field, have compressed their matter to such a high pressure that the sole electrostatic force of the electrons can no longer halt the gravitational collapse. The wavefunction anti-symmetry prevents them from being squeezed into the same place at the same time, that is, to occupy the same quantum state, which stops the collapse. Therefore, a white dwarf does not shrink further into a black hole because of the so-called *'degeneracy pressure'*, which is simply that pressure emerging due to the fermions' resistance to occupying the same quantum state.

Fig. 133 White dwarf of Sirius A and the Crab Nebulae containing a neutron star.

In neutron stars, the pressure is so high that protons and electrons melt into neutrons; yet, still, the degeneracy pressure can counteract the gravitational collapse. The degeneracy pressure is so high, it becomes an order of 20 magnitudes (100 billion of billion times) harder than a diamond. Finally, in the case of a black hole, the gravitational field becomes so strong, it overcomes even the degeneracy pressure and the star definitely collapses into an object that present physics is not advanced enough to describe.

Therefore, bosons and fermions behave very differently, not only in pairs but more generally in composing bodies with larger numbers of particles. For this reason, the statistical and mathematical tools that describe the former or latter particles are different as well; they are called the *Bose-Einstein statistics* and *Fermi statistics*, respectively.

The fermionic anti-symmetry also tells us something more general, beyond the fact that fermions are not allowed to occupy the same place in space. Because the position of a quantum particle is considered just one of the quantum states that describe particles, it is natural to extend this, conjecturing that fermions cannot be in the same quantum state in general. For example, two fermions can't occupy the same energy level or spin state at the same time, just as they can't be in the same place. More precisely, this idea was put forth in 1925, in the famous principle of the Austrian physicist Wolfgang Pauli, who summarized it as follows.

Fig. 134 Wolfgang Pauli (1900-1958)

Pauli exclusion principle: *"Two identical fermions cannot occupy the same quantum state simultaneously."*

It should be clear at this point why this is equivalent to stating that *"the wave function for a system of two identical fermions is exchange anti-symmetric."*

This also has very important implications in terms of the structure of atoms and the reasons why the periodic table of elements has that structure and is just as it is. As every chemist knows, Pauli's exclusion principle is at the base of the structure of the atomic table of elements. It dictates the buildup of the electrons in the atomic orbitals. Two electrons with the same spin are not allowed to occupy the same atomic energy levels, the same orbitals (or orbits, if you still think in terms of Bohr's atom). If an orbital is already occupied by two electrons with opposite spin, a new electron can only pile up at a higher energy level in a different orbital shell. It is according to this principle that the atoms' structure and their chemical properties come into being. Fig. 135 illustrates this with spin states.

The upper row shows the correct ordering: only one electron with spin-up or only one with spin-down or both occupying the same energy level but with opposite spins. The lower row instead shows what Pauli's exclusion principle *does not allow* for: two electrons with the same spin in the same quantum state. What it does not allow for are electrons with both spin-up or both spin-down. Nor does it allow for more than two electrons in the same quantum state. (The latter is impossible because there are only two spin states.)

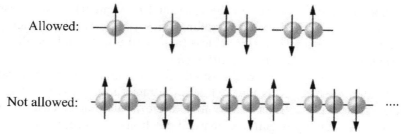

Allowed:

Not allowed:

Fig. 135 Pauli's exclusion principle illustrated with spin states.

To summarize, matter is 'hard' not because 'marbles' are hitting each other but because of the mutual interaction force fields between the particles, the presence of electrostatic repulsion and the exchange anti-symmetry of fermions. These account for the features of the entire material Universe we observe and perceive.

2. Why is matter 'stable'?

For several reasons, we have focused our attention on the Pauli exclusion principle. The first reason is that it is one of the most fundamental principles of QM and because it is at the very base of the periodic table – that is, of the structure of matter and all of chemistry. A second reason is that it answers – or, more precisely, is one of the answers to – another important question: Why is matter stable? Once again, a seemingly silly question.

However, consider that all atoms, which form the molecules and complex material structures of which we are made, have a nucleus which is made of electrically positively charged protons and neutrons (though, because they are electrically neutral, these do not contribute to the overall nuclear electric charge) and around which negatively charged electrons form the inner and outer shells of the atom's electron cloud. As you learned in school, electric charges of opposite signs attract each other. So, the question is: Why do atoms not collapse onto themselves and form a sort of proton and electron mixture? Why doesn't the Universe collapse into a neutrally charged and much smaller chunk of inert matter?

We know that electrons do not reside in the nucleus because we can easily extract them – that is, ionize the atom – with much less energy than would be necessary to extract electrons if they resided inside a nucleus. Moreover, let us not forget how the good old Bohr model, in which he conceived of electrons as orbiting the nucleus, is not tenable, not even in classical physics. It is well-known that an electric charge that is accelerated (an electron orbiting a nucleus is permanently deviated, that is, it suffers from a centripetal acceleration due to the rotational movement) must necessarily radiate EM waves and would therefore quickly lose its orbital energy and collapse onto the nucleus nevertheless (a general effect valid for all electric charges, called the '*Larmor radiation effect*').

However, one immediate, almost intuitive possible answer to the stability of the atomic structure is that, while an attraction does exist between the nucleus and the electrons of an atom, there is also an electric repulsion between the equally negatively charged electrons because, as we all know from school, electric charges with the same sign will repel each other, as is pictorially shown in Fig. 136. So, we must expect that some electrostatic equilibrium exists between the attractive forces, which tend to make the electrons fall onto the nucleus, and the repulsive ones, which tend to separate them.

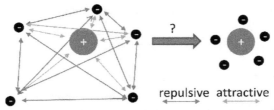

Fig. 136 Proton-electron avoided collapse due to the electrons mutual electrostatic repulsion.

Moreover, we shouldn't forget the already-mentioned function of the strong nuclear force. Because protons have the same positive electric charge, they repel each other, as shown in Fig. 137 left, and all the nuclei they form with the neutrons should evaporate quickly. The atomic nucleus has only protons and neutrons but no negative electric charge that could keep the protons bonded. Fortunately, Nature has provided for the existence of the strong nuclear force which acts only at very short nuclear distances (which is why we have no experience with it in our everyday lives) and with the opposite attractive force between protons and protons as that between protons and neutrons or among neutrons themselves, as shown in Fig. 137 right.

So, this is, in part, the answer and solution to the problem of why matter is stable. However, it can't be the whole story. First, it can be shown that, if we take into account only electric forces, atoms must be much smaller than their actual size. No, we would not notice this using some standard ruler, as every object like a ruler would also shrink and nobody would notice a difference in length. (Think about that!) However, the atomic spectra would be completely different from those we observe and nowadays can calculate using the Schrödinger equation.

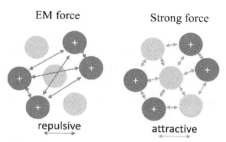

EM force   Strong force

repulsive   attractive

Fig. 137 Proton-proton EM interaction (left) and proton-proton as proton-neutron strong force interaction (right).

Yet apart from these theoretical calculations, there is a very simple fact which tells us that there must be much more than mere electric repulsion effects that avoid an atomic collapse. In fact, what about the hydrogen atom? Here we have only one positively charged proton and a single negatively charged electron. The repulsing effect of other electrons is missing. And yet we know that hydrogen is a stable atom; no collapse of the electron on the nucleus occurs.

Therefore, we must finally recognize that, among many other things, classical arguments of physics are not able to explain – or, at least, cannot sufficiently explain – the stability of matter as we know it. We must resort to something else. Yes, you guessed it: quantum mechanical explanations.

The first step is to switch from classical deterministic physics and Bohr's atom to Heisenberg's uncertainty principle and orbitals. These can help us understand why a collapse of well-defined point particles on a central nucleus in an atom cannot occur simply because there are no such things as point particles in the first place. In this conception, there are only electron clouds – or, more precisely, probability density clouds – that no longer have something in common with the idea of electrons orbiting along deterministic paths around a central nucleus. The root of the 'fuzziness' of this model is the uncertainty principle. An electron can't collapse onto a limited, almost point-like region because then its uncertainty in position (recall the Δx in Heisenberg's formula, which you should know by now) would be almost zero. This, in turn, would increase the uncertainty over its moment (Δp), which means over its speed, and, in turn, implies its kinetic energy, making it jump out quickly from the nucleus again. To maintain its limited indeterminacy over the momentum, Heisenberg's formula tells us that it is forced to maintain its whereabouts at some distance from the nucleus.

Therefore, to the electrostatic equilibrium, we must also add a sort of 'equilibrium of uncertainty', according to which a tradeoff between the uncertainty of the position and that over the momentum must be realized.

So, we are nearing a solution to the problem. What prevents atoms from collapsing onto themselves is not only a force of repulsion, attraction or any other sort of force field but a fundamental principle of the natural world. Electrons, like any other particle, have no intrinsically defined position or momentum in space. They are, in themselves, fuzzy entities that can't fall onto a nucleus because that would localize them too much, against Heisenberg's uncertainty principle, as any attempt at localization implies the enhancement of its momentum.

Notice that this further confirms something very fundamental about Heisenberg's uncertainty principle itself. We already saw, and now we find further reasons to confirm, that this principle can't be intended as a limit on our knowledge of the state of the system alone. The system simply does not have any 'state' in the classical sense of the word. Heisenberg's uncertainty principle does not simply tell us that we can't determine the position and momentum of a particle at the same time because this tends to imply misleadingly that it is all about our ignorance of the position and momentum of the electron. Heisenberg's principle tells us that there are no such things as positions and momenta in the first place before we measure it; these are only fictions of our minds accustomed to living in a classical, Aristotelian world. Objects in the quantum world are intrinsically fuzzy with or without our measuring them. If we would not interpret Heisenberg's principle in this way, the stability of the hydrogen orbital, like that of any other atomic orbital, could not be explained. Atoms do not collapse or remain stable because humans are ignorant of some facts but because Nature is intrinsically indeterministic. (Proponents of the de Broglie-Bohm pilot wave theory or of Everett's Many Worlds Interpretation, who still believe that determinism can be saved, would not agree on these latter statements; we will take the interpretations of QP in the second volume.)

But, again, something is still missing. Calculations show that electrostatic considerations, together with quantum effects, indeed explain why atoms do not collapse but still this is insufficient for explaining the world we observe. In fact, what remains unexplained is why the electrons arrange themselves around the nucleus in just that way we observe, and which is eloquently represented in the Mendeleev periodic table of elements. The physical and chemical proprieties of the elements, with their spectra and energy states, is very peculiar, not just a random and linearly ordered intuitive system. Every time we add a proton to the nucleus and an electron around it, its energy states, its electric field, its chemical behavior, etc. do not at all fit into a model which is limited to uncertainty and classical electric effects. A third ingredient must come into play.

This is, of course, Pauli's exclusion principle. In the previous section, we analyzed the deeper meaning of it. Around the volume of an atomic nucleus,

Pauli's exclusion principle forbids the coexistence of two or more identical fermions (here, in our case, the electrons) in the same quantum state. This means that no more than two electrons can occupy the same space in the atomic cloud. Two, because they can differ in the spin orientation (which means they are not identical in their quantum state). Unlike bosons, which are not subjected to Pauli's principle, electrons instead must pile up in the atomic potential. No more than two electrons can be placed in the same orbital lobe. The energetic hierarchy of the orbital filling of an atom must obey Pauli's exclusion principle.

So, in summary, the stability and structure of matter have four causes:

1) the electrons' electrostatic repulsion,
2) the strong nuclear forces,
3) Heisenberg's uncertainty principle,
4) Pauli's exclusion principle.

When physicists took into account these forces and principles and made the appropriate calculations (first and foremost with Schrödinger's equation), everything corresponded nicely with the size and physical-chemical properties of the atomic world we observe. Though the calculations necessary to obtain the spectra of atoms with high atomic numbers are very complicated and sometimes require numerical integrations, one might nevertheless say that this is yet another triumph of the power of mathematics.

One important takeaway message of this chapter is that without quantum effects, the world we know would immediately collapse onto itself. In addition, and we hope you agree, that tiny signature change characterizing the fermions' wavefunction after the 360° rotation or appearing due to exchange-symmetry is the most important minus sign in the Universe!

The table on the next page schematically summarizes what we have found so far.

Bosons	Fermions
Photons W^+ , W^- , Z^0 Gluons Higgs Gravitons	Electrons Protons Neutrons Neutrinos, etc.
Integer spin $0, \pm 1\hbar, \pm 2\hbar, \pm 3\hbar \ldots$	Half-integer spin $\pm \frac{1}{2}\hbar., \pm \frac{3}{2}\hbar, \pm \frac{5}{2}\hbar, \ldots$
Rotational and exchange symmetry 	Rotational and exchange anti-symmetry
Singlet state $\lvert \Psi \rangle = \dfrac{\lvert + \rangle \lvert - \rangle + \lvert - \rangle \lvert + \rangle}{\sqrt{2}}$	Singlet state $\lvert \Psi \rangle = \dfrac{\lvert + \rangle \lvert - \rangle - \lvert - \rangle \lvert + \rangle}{\sqrt{2}}$
Symmetric wavefunction 	Anti-symmetric wavefunction

3. Phase matters: the Aharonov–Bohm effect

In the example of the neutron interferometer in chapter III.4, we saw how a magnetic field can influence the wavefunction of matter waves by inducing a phase shift on it (see Fig. 83 and Fig. 84). In 1959, Yakir Aharonov and David Bohm, an Israeli physicist and an American physicist, respectively, predicted that magnetic fields can do even more than that. The detailed description and analysis of the physics standing beyond this phenomenon and its experimental verification is quite long and would require mathematical subtleties we will avoid here. (However, the interested reader can find a more in-depth description by the author in the form of a video lecture [14].)

As we know, particles are described by wavefunctions or wave packets. Also, we know how, due to the wave-particle duality, the particles diffracted by the double slit and revealed at the detector screen must be described by an incoming plane wave which is diffracted at the slits and interferes on the screen, forming the interference fringes. That is, the phenomenon must be described by two wavefunctions, Ψ_1 and Ψ_2, each representing the particle going through slit one and slit two, respectively. (Recall Fig. 8.) Now let us consider not just electrically neutral particles, such as photons or neutrons, but charged particles – say, electrons – going through the two slits. So far, it has been implicitly assumed that, in the space between the two slits and the detecting screen, no electric and no magnetic fields are present. One might ask whether and how things will change with charged particles if we turn on a magnetic field. The easy answer is that everything will change for the simple reason that magnetic fields deviate charged particles. This, of course, will dramatically change the interference pattern or even completely destroy it.

However, in the physical formalism of QM, it turns out that, if the magnetic field is bound inside a limited spatial region outside the particles' path, there will nevertheless be a phase shift on the wavefunctions of the two particles. This will change the observed diffraction pattern. That is, without direct contact between the particles and the magnetic field, but simply because of its existence nearby, the particles will 'feel' its presence and react accordingly.

To illustrate this effect, let us first explain how relatively easy it is to build a magnetic field inside a small region of space, but without any present outside. This can be done using the good old electromagnet 'solenoid', that is, a cylindrical coil of electric wire (Fig. 138 left) which, when carrying a current i, generates a nearly uniform magnetic field inside and similar to that of a bar magnet outside (Fig. 138 right). If the solenoid is long enough (ideally, infinitely long), all its magnetic field lines will be stretched out and remain confined in the coil's inside region.

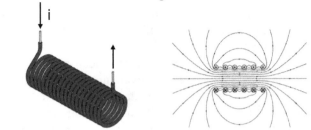

Fig. 138 The solenoid (left) and its magnetic field (right).

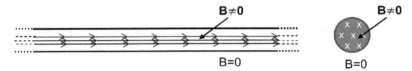

*Fig. 139 The long solenoid's magnetic field (**B**):*
a uniform field inside (left, cross-sectional view) and none outside (top view right).

Fig. 139 left shows the solenoid's field lines (a magnetic field is usually labeled with the vector **B**): outside, there is none (**B**=0) whereas inside, the field lines are ideally perfectly parallel and uniform ($B \neq 0$). Fig. 139 right shows the top view of the solenoid; a well-defined region has been created with a uniform magnetic field inside and none outside. (The white crosses represent the presence of the field pointing inside the diagram plane.)

Now consider the double slit experiment again, in this case with charged particles, such as electrons. As usual, we have a situation like that represented by Fig. 140 left, with one or more electrons traveling through the two slits and producing the usual interference pattern on the screen. Note that, in the picture, we take first the 'semi-classical' point of view. The two electrons' wavefunctions – $\Psi_1(x)$ and $\Psi_2(x)$ – label the two possible paths it can take. Because these have different lengths, they also determine a phase difference at point P. That is, we think in terms of well-defined trajectories of point-like particles instead of imaging propagating waves (something one should not do if one cares for the correct ontological interpretation).

The two wavefunctions interfere on the screen and will produce an interference pattern according to the phase difference induced by their mutual path difference, which depends on which point on the screen is considered. In principle, it is the same experiment as Young's double slits experiment (other than, of course, the fact that we are using electrons instead of photons).

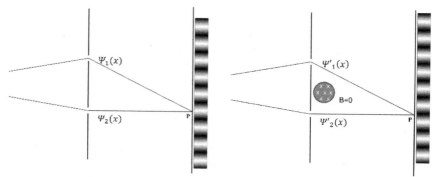

Fig. 140 The Aharonov–Bohm effect: the wavefunction 'feels' a magnetic field even without a direct contact.

Now consider that we place behind the slits a magnetic solenoid, as in Fig. 140 right. More precisely, this is a very long solenoid such that its magnetic lines are present only inside it, but no external magnetic field exists with which the particle will get in physical contact. This means that if we look at things from the classical point of view, in which we imagine the particle going through the upper OR lower slit and following the well-defined upper OR lower path towards point P, they will not traverse the solenoid's internal field and we will not expect them to be affected at all. Thus, the two situations represented by the left and right sides of Fig. 140 should not differ; we expect the particles to behave in the same manner because, having no apparent physical contact with the solenoid's magnetic field, their physical state is not expected to change and there is no reason to believe that the interference pattern should change or be affected.

This experiment was conducted in the 1960s, and guess what? If the electric current in the solenoid is turned on and the magnetic field builds up, the interference pattern also changes! This looks like the particle going through one or the other slit 'knows' that a field exists inside the cylindrical coil, even without 'seeing' it, and behaves accordingly. The interference pattern is shifted according to the strengths of the magnetic field inside the solenoid, even if the classical path of the particles never crosses the solenoid and always travels in a region with zero magnetic field. Again, some sort of 'spooky action at a distance' seems to be at work here.

This is a purely quantum effect because, for the electron going through the slits, neither its path nor its energy or momentum changes; it does not experience any sort of force. It can't in the absence of any external influence and magnetic or electric fields. Technically, what happens is simply that the phase of the two wavefunctions Ψ_1 and Ψ_2 changes by turning on the field inside the solenoid. That is $\Psi_1 \to \Psi'_1$ and $\Psi_2 \to \Psi'_2$ with the primed wavefunctions differing only by a phase from the unprimed ones. What changes is, therefore, the probability wave's phase, something that in classical physics has no meaning and which is only a quantum concept. However, as you know, any phase change between two waves coming from the first and the second slits interfering on the screen plane will produce a different interference pattern; that is, it shifts the white (or black) fringes in one direction or another along the detection screen. (For illustrative purposes only, in Fig. 140 the bright fringes have been replaced with the dark ones.)

All this might lead to some speculation about what really happens here. It seems that, in QP, particles have a sort of holistic tendency to not only perceive the presence of a field in their neighborhood but to also be somehow directly or indirectly connected to large parts of the environment that surrounds them. They seem to 'feel' instantly, in some holistic way, how the experimental setup is made as a whole. Indeed, never forget that the correct way to look at the double slit experiment in QM is the one shown in Fig. 141. This, again, should remind us that the notion of a classical trajectory is misplaced. We must think of particles not as particles at all but as propagating matter waves.

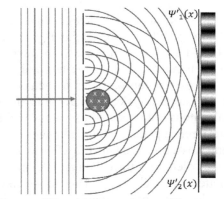

Fig. 141 Wave-like point of view of Fig. 140.

We should always think 'dually' in terms of particles and waves, with the latter diffracted in all directions as circular wavefronts towards the detecting screen. From this perspective, it comes as no surprise that the physical objects we call 'particles' are somehow 'informed' by the existence of a magnetic field in the solenoid. Their wavefunction naturally encloses all the space existing between the two slits wall and the screen; therefore, they will also come into contact with the solenoid. The Aharonov-Bohm effect could be intended as a natural consequence of the ubiquity of the wavefunction, which nevertheless collapses as a single point on the screen at the time of detection.

Yet things are not as simple as that. (Well, they're already weird enough; in QP, nothing is trivial.) This is because there is another possible interpretation of what is going on here. This effect was not found simply by a serendipitous experimental discovery but was already predicted by Aharonov and Bohm using the mathematical formalism of modern physics, which defines a vector quantity (a physical abstract quantity with a magnitude and direction, like the angular momentum, the spin, the velocity, etc.) that is called the 'magnetic vector potential', or 'vector potential' in short, usually labelled with the letter "A". It is a vector field that extends the definition of the electrostatic potential E to a time-varying one, that is, to EM fields. The vector potential also defines the magnetic field B itself. A rigorous description would require some mathematical stuff (for those interested in the details, again, see [14]) but, intuitively, we might visualise it as a field which gives rise to the magnetic field if it is twisted or rotated. We might say that any magnetic field specifies the amount of rotation of the vector potential field lines. The vector potential is an extension of the notion of the electric and magnetic fields, unifying them into another field that contains them both as $A(E,B)$. It is simply an abstract mathematical tool that describes EM fields, or at least this is what one might believe before the Aharonov-Bohm effect. The point is that the vector potential is not necessarily zero in the absence of a magnetic field. For an absent magnetic field, it simply becomes a constant but non-zero quantity. One says that the field is 'irrotational' because the field lines of A are straight; they are not affected by 'vorticity'. However, the vector potential exists. This is precisely what happens in the surroundings of the solenoid: The magnetic field outside it is zero but the vector potential remains and could affect the wavefunctions' phase. It can be shown that, according to QM, the wavefunctions' phase is changed by the presence of the vector potential along the paths of Ψ_1 and Ψ_2. The phase difference between the two wavefunctions is determined not only by the different paths' lengths but also by the strength of the vector potential that the particles 'feel' along their path to the screen. (More precisely, the induced phase difference is proportional to the enclosed magnetic flux of the two paths, regardless of whether they cross the magnetic field.) This effect shows that we can change the phase of a wavefunction's particle without the presence of an electric or magnetic field, but solely by the change of the vector potential.

Yet here, again, we confront the fact that what was introduced as a mere mathematical construct, intended to serve as a simple generalization of the electric and magnetic fields, is actually an observable, physically real entity. Therefore, we should not regard the vector potential as a purely abstract mathematical extension, but rather as a real field that exists in the physical world, as the electric and magnetic fields do. In classical physics, the notion of a field's potential is never something you can measure directly. In classical physics, what you always measure are forces, which are the gradient of the potential (the amount of variation in space of the potential), not the potential itself. Whereas, in QM, the vector potential is something real that can be directly measured by its action on the phase of particles. As in the case of the interpretation and meaning of the wavefunction, here arises the question: To what extent are the mathematical objects we employ in QM real or abstractions?

This is another nice example of how QM is mathematically and formally perfectly consistent. Calculations make everything clear and well-defined in its formal development. These calculations, indeed, make precise predictions that are verified in the laboratory; in this sense, no doubts or difficulties exist. Physics is, first of all, about models, calculations and empirical verification. Physicists are not asked to furnish deeper philosophical interpretations. This is why so many physicists will tell you to not bother about philosophical questions and to "shut up and calculate". Yet here, again, from the philosophical perspective and the interpretational point of view, a deep sense of dissatisfaction remains, at least for many, and for the author, too. That's why Bohr was happy with QM as it is, unlike Einstein, who longed for insights that appeal to our intuition. Yet it remains doubtful whether this attitude – which essentially negates the innate human nature that seeks to understand and wants to know more – will survive the test of time.

4. Path integrals and Feynman diagrams

This section will try to briefly convey what the so-called '*path integral formulation*' of QM is and detail the related Feynman diagrams of particles' scattering processes. We can take only a quick and superficial look at it, on the fly, because describing this topic would require lots of complicated mathematical tools that go far beyond the aim of this book. However, this information is worth mentioning because it contains some important messages which clarify the differences between a classical view and a quantum view and the understanding of the physical world from the perspective of QFT.

Fig. 142 Richard Feynman (1918-1988)

The path integral formulation is due to the efforts of the famous American physicist Richard Feynman and represents a deeper description of QM that empowered physicists to extend it to relativistic QFT

– more precisely, to quantum electrodynamics (QED), which is the relativistic version of QT that considers EM interactions only and that gave birth to the modern standard model of particle physics. Essentially, the path integral formulation gives us a new point of view regarding the trajectory in space and time of a particle.

Consider the classical Newtonian trajectory of a particle (see the straight line in Fig. 143) from an initial point x_i at time t_i, which we label with space-time coordinate (x_i, t_i), to another final point x_f at time t_f, labeled (x_f, t_f). The space variable x is illustrated here with only one dimension but it can also be thought of as a two- or real three-dimensional space (and eventually also an n-dimensional one).

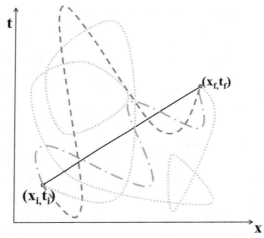

Fig. 143 A straight line connecting two points in space-time and other alternative paths.

Notice that in this representation, on the horizontal axis, the abscissa, we have a space coordinate, while the vertical axis contains the time parameter. This means that a classical particle or body that is not subjected to any external force will proceed along a straight line with a slope depending on its speed. (A particle with no speed traces a vertical line, while one with a higher speed will make the line tend towards the horizontal direction.)

In classical physics, it is a well-known fact that, when you kick a body and then leave it free to move without additional action from external forces (such as friction), it will move along a straight line forever. This is the shortest path between two points, also in this space-time depiction. This is simply an everyday common experience: The shortest path between two points is the straight line – a piece of intuitive wisdom since the times of the ancient Greece philosophers, like Euclid. who formalized this in mathematically precise terms in what we today know as '*Euclidean geometry'*. However, there seems to be no apparent reason for that. Why does Nature always choose the straight line? Why not the dashed, dotted or dashed-dot path of Fig. 143? Again, another apparently silly question, but it turns out that, in QM, there is a very special reason for this.

In the section about the state vector, Schrödinger's equation and orbitals, you might recall that we quickly introduced the Dirac notation. There (see Eq. 11), we described the state vector as $|\Psi\rangle = c_1|e_1\rangle + c_2|e_2\rangle + \cdots$ with e_1, e_2, etc. being the possible discrete eigenstates in which the system can be found after the measurement. Then (see Eq. 13), the probability that the system will be in the i-th eigenstate is given by the modulus square of the i-th coefficient, which in Dirac notation is the projection of the state vector onto the i-th basis (the amplitude) in squared modulus, that is: $|\langle e_i|\psi\rangle|^2 = |c_i|^2$. By this expression we mean the transition probability that the system undergoes a transition from state $|\Psi\rangle$ to $|e_i\rangle$ (see also Eq. 18).

Now, here, we have something similar but applied to the continuous case of a particle traveling in space and going through all possible paths from an initial state to a final state. (Recall, also, that positions in space and time are 'states'.) In QM, the probability amplitude for a transition from an initial position x_i at time t_i to a final position x_f at time t_f, is described by a '*propagator*': $\langle x_i,t_i|x_f,t_f\rangle$. (This is a probability amplitude, not the modulus square, for the sake of simplicity and conformity with the standard notation you will find throughout textbooks.) With the propagator of a particle, one does not calculate the trajectory from a point in space-time (x_i,t_i) to another point (x_f,t_f), but the probability that a particle initially in (x_i,t_i) will be found later in (x_f,t_f). To put it in more prosaic terms: "Give me the initial state of a particle and, by calculating the modulus square of the amplitude of the propagator, I will tell you the probability of finding it in a specific final state".

Feynman found that, in QM, this kind of expression is the most adequate one. Without going into the mathematical details, one can intuitively state his approach as follows. If you want to know the probability that a point particle at some time with a specific initial position will be found at another specific final state in space and time, you don't consider only the straight path. Rather, you consider all the possible trajectories leading from (x_i,t_i) to (x_f,t_f). (The trajectories in Fig. 143 are only three of the infinite possible ones.) This should not come as a surprise because, as we know all too well by now, in QP nothing is determinate and well-defined. There are no classical paths, and the particles' trajectories should be considered fuzzy because of Heisenberg's uncertainty principle and the wave-particle duality. Therefore, the straight line should be considered fuzzy – that is, it may not be straight at all. Moreover, the superposition principle governs QM; therefore, it is perfectly possible to have a particle which is simultaneously in several states and here is represented by a wavefunction containing all possible paths between two points, though at the instant of the measurement act, it displays only one of these states. Finally, we also saw that because of the time-energy uncertainty relation, the instant in time when a specific phenomenon occurs is never precisely determined. It depends on the system's energy. Thus, it is a quite natural thing to consider all the infinitely possible trajectories, from one space-time point to another.

Therefore, Feynman conceived of any possible path as being equally probable, just as the straight-line path. He then represented the propagator as an integral over all these equally probable paths, though each had a different phase factor because different paths lead to different relative phases of the wavefunctions. Think of the two wavefunctions in the double slit experiment. On that occasion, we had only two paths, whereas here we calculate over all the possible trajectories. Another way to see this is to consider Young's experiment with an infinite number of slits. In some sense, we might say that this situation is a natural generalization of the two slits to an infinite number of slits experiment. With an increasing number of slits, you must add more amplitudes. Finally, when you have drilled so many holes and slits, you will no longer have the slits' interface itself. You will obtain the natural situation of a particle traveling from an initial point to a final point.

The probability is therefore calculated over the infinitely possible set of paths from an initial position to a final position, but with each of these paths having a phase factor of its own. This phase factor weights every path, that is, not only the straight line but all the possible, more or less intricate and imaginable paths connecting the two points. The integral sums them all together, hence the name 'path integral'. One speaks also of the probability amplitude as a *sum over histories* where, obviously, the single 'history' is represented by the single path.

Guess what happens when all the possible paths with all their phases are considered and interfere with each other? Due to the interference effects which are summed up, only the straight-line path has constructive interference and 'survives'. Destructive interference cancels out all the others.

So, the reason why particles move along classical trajectories is rooted in quantum interference phenomena. The classical understanding of the deterministic and definite path of a particle, that kind of trajectory we have in our minds intuitively from classical mechanics, is the sum of all the infinitely possible trajectories, plus their relative phases considered. Feynman's path integral formulation of QM conceives of a straight line as the sum over all 'random walks'. However, while all the wavefunctions representing the curved paths delete each other due to destructive interference, those which are very near to the straight line do so only partially. 'Near' means at the microscopic level, though other paths are allowed. That is why, in QM, everything is so fuzzy. Nearness is determined by the value of Planck's constant. If $\hbar$ were zero, we would recover a perfectly straight line, that is, a classical world. However, because $\hbar$ is tiny but not zero, at a microscopic level, particles can also follow slightly non-straight paths, even in the absence of external forces, or more generally not classical trajectories.

This is a more equivalent way to formulate QM than the standard Dirac one, and the eigenequation formalism with energy eigenvalues and Schrödinger's equation. Path integrals play an important role in modern QFT and particle physics. The most important application of Feynman path integrals can be found in the calculation of the probability of scattering processes which allow for the quantification of the interactions between particles in particle accelerators.

In fact, closely connected to the path integral formulation are the 'Feynman diagrams'. These are visualizations of particles scattering and their decay processes.

For example, the diagram in Fig. 144 left depicts an electron-positron annihilation. Recall that the positron is the anti-particle of the electron – that is, it has the same mass but an opposite charge and spin. When the two approach each other, they annihilate emitting gamma-ray photons. As in the previous diagram, we take the convention of the horizontal axis as space and the vertical axis as representing time flow. This shows how the two particles near each other, at a certain instant, interact in some region of space and time, exchange some other particle (not defined herein) and then transform into two high-energy gamma photons. The propagators describe this interaction region. Note that the path arrows' directions show the direction of time. As a mathematical artifact, for anti-particles, time is labeled as going backwards. This might be somewhat confusing at first: The direction of the antiparticle's movement in the diagram must be read as going upwards!

Fig. 144 right shows another interaction, namely, an electron-electron interaction. However, this time, the horizontal direction is showing the flow of time and the vertical direction is showing the particle's space position. (Both conventions are illustrated, so you can become acquainted with additional literature which uses them both interchangeably.) Two electrons arrive from the left, interact during an exchange of a virtual photon, which implies an exchange of momentum, and thereafter continue along two deviated paths in opposite directions. This quantum field theoretical perspective represents a well-known fact: Electric charges with the same sign repel each other.

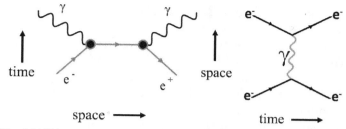

Fig. 144 Electron-positron annihilation and electron-electron scattering
in terms of Feynman diagrams.

Recall, from the section on the zero-point energy and virtual particles, that the virtual photon, like any virtual particle in general, is never something which can be observed experimentally. Virtual particles are called 'virtual' simply because they are not directly observable objects. In scattering processes which employ Feynman diagrams, these are more calculating tools whose graphical depiction helps our intuition, but should not be considered an objective description of the way things are really. The only things we can directly observe are the scattered electrons. No experiment can ever detect virtual particles in isolation, not even in principle. This is because the time-energy uncertainty relation limits their existence. Therefore, because they are so extremely short-lived, even if they travel with the speed of light, they will never reach a detector.

Feynman diagrams can be much more complicated. Fig. 145 shows an example of one of the many processes that can occur when two protons interact (time direction is horizontal again).

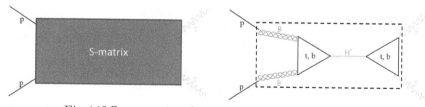

Fig. 145 Representation of a proton-proton scattering process:
an S-matrix block diagram (left) and its expansion in Feynman diagrams (right).

Coming from the left, when something is approached, a scattering process occurs and, subsequently, a couple of photons emerge. In the laboratory we can observe only the incoming protons and the emergent photons; we can only infer indirectly what happens in between. In fact, lots of complicated things can happen during the scattering process inside this region (box of Fig. 145 left). In QFT, this is described as the 'scattering matrix', or in short, the 'S-matrix', which, thanks to the mathematical procedure that Feynman introduced, can be calculated and reveal the internal structure of the scattering process.

This shows how the famous Higgs boson was discovered at the LHC in 2012. The two protons emit two gluons, the bosons responsible for the exchange of strong nuclear forces, which fuse into a top-quark loop, which itself couples with the Higgs field that interacts with another quark loop and decays into the two photons. As you can see, it is quite intricate. Regard it simply as an example which has only illustrative purposes and which shows how things can be quite complicated in the microcosm of interacting particles.

It is, however, not surprising that the scattering between two protons is more complicated than that between two electrons. This is because electrons, as far as we know until today (at least until 2019), are really elementary objects—that is, we consider them to be point-like objects with no composite structure and which are sensitive only to the EM force and the weak nuclear force, but that remain unaffected by the strong nuclear force. For them, a virtual photon is sufficient to make them scatter. Not so for protons. These are composite objects made up of quarks and, thereby, having a complex internal structure inside which complicated nuclear strong and weak force processes are dominant, with the strong force responsible for keeping each quark bonded with each other and to form the particle we call 'proton'.

These Feynman diagrams emerge from a mathematical procedure: the *'perturbative expansion'* in several terms of the S-matrix. Again, it is about the calculation of an amplitude – here, the scattering amplitude, which means the probability that a system goes from an initial state $\langle i|$ to a final state $|f\rangle$ as $\langle f|S|i\rangle$ (a generalization of the path integral), where S is the scattering matrix. Then, in perturbation theory, the S-matrix is represented by an infinite power series, that is, a sum like $\langle f|S|i\rangle = \sum_{n=0}^{\infty} S^{(n)}$, with $S^{(n)} = \sum_{j=0}^{n} S^{(j)}$ again another sum over what is pictorially represented as the Feynman diagrams.

If that confuses you, don't worry. Calculating Feynman diagrams is a complicated science. Just imagine this pictorially through a general scattering process between two incoming, interacting and outcoming particles like in Fig. 146 left. This can be represented as the sum over a large number of more elementary scattering processes. The calculation involves the scattering process at higher orders which, from a purely formal and mathematical point of view, is the sum over many different scatterings.

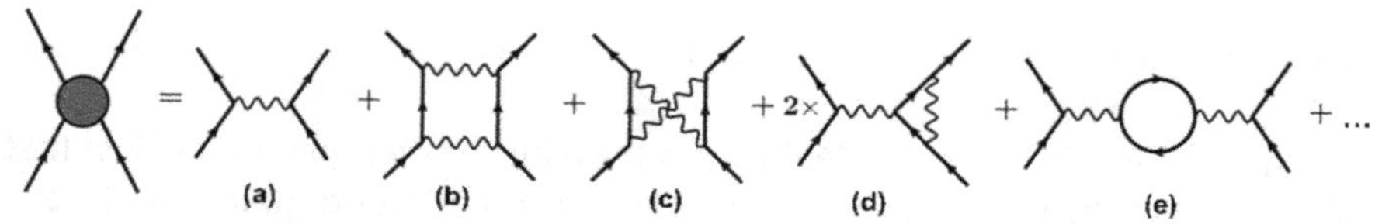

Fig. 146 Perturbative expansion in Feynman diagrams of a scattering process in QED.

It is like the sum of many elementary waves which gives, as a result, a single larger wave. Similarly, here, one can have a one- or two-photon scattering ((a), (b) and (c), respectively) plus a photon emission which interacts with the other particle, the latter contributing a self-interacting process (d) plus a one-loop term, that is, a photon that decays in a particle and anti-particle but shortly after annihilates again (e) and so on and so forth for an infinite number of terms.

So, as in the case of the path integral formulation of the path of a particle from one point to another as the sum over all possible paths (the sum over histories), here we have the analogy of the scattering process as a transition amplitude made of the sum over all possible scattering processes. These interactions and scatterings also involve virtual particles. In QFT, curiously, at short distances, one must also consider these strange self-interactions of particles with itself. If one did not do so, the theory would be 'non-renormalizable', that is, it would exhibit annoying unphysical infinities.

The question at this point is the same as the one we already encountered in several other situations. From a mathematical point of view, QT again reveals itself to be perfectly consistent. However, we might have some doubts as to whether these diagrams actually represent a phenomenon with some ontology – that is, if it has anything to do with a reality that the diagrams seem to suggest. As we said, these diagrams are only terms in a long series of a sum, not the sum itself. It is like describing a particle at rest as the sum of two equal particles moving in opposite directions. That would be mathematically consistent and, if checked experimentally, would be confirmed. However, it would not have much to do with physical reality. Feynman diagrams have been invented for mathematical convenience, as otherwise, the math would become very intricate and difficult to manage. This convention works wonderfully. It has given us the standard model of particle physics, which is one of the best experimentally tested theories of science. However, we should be aware of its limits when trying to imagine and interpret the microscopic world, making inferences about its ontology.

5. The quantum Zeno effect

It may be useful to conclude this chapter with another quantum effect that did not attract the same amount of attention as other weird quantum effects, but it is worth mentioning and keeping in mind that it exists. Another interesting quantum effect is the '*quantum Zeno effect*'. The quantum Zeno effect is a phenomenon in which repeated and fast measurements slow down a quantum system's time evolution. It is a peculiar quantum phenomenon and applies to every system and situation in which any kind of transitions occur in time. To understand this, it might be useful to compare it first to our everyday experience.

As is well-known, if you sufficiently heat a pot of water, it will sooner or later start to boil. That is, the water pot is a system undergoing an energetic thermodynamic transition from non-boiling water at a low temperature to boiling water at a higher temperature (100°C at sea level). It goes without saying that this is supposed to happen regardless of whether or not you are watching the pot.

Not so in QM. In this case, we are dealing with a microscopic and quantum system at the atomic level or not much more than that. The Zeno effect describes the dynamics in the quantum regime whereby the transition probability from a quantum state to another is modified because of the measurement – that is, simply because it is 'observed'.

For example, take an unstable atom that has some probability of decaying next in time. The radioactive '*half-life time*' of a specific type of atom is the average time after which 50% of a large number of atoms of the same type has decayed. This half-life time for each atom is not something we can engineer. It is a constant given by the laws of Nature. However, the same atom, if observed continuously, will never decay. This implies that the probability that an atom will undergo an energy transition emitting a photon is lowered if the system is subjected to frequent ideal measurements. Consider an atom that is excited by a photon and that undergoes an energy transition to a higher excited state. It will, sooner or later, relax to the ground state and emit a photon. The probability that this occurs within a specific time interval is entirely dictated by quantum laws. If, however, the atom is subjected to fast and repeated measurements, the probability that this transition occurs will become lower and lower, as the frequency of these measurements increases.

Ideally, if the measurements are so fast that we can consider the observation as continuous, the transition probability tends to zero and we can consider the system almost frozen in its eigenstate. In other words, measurements can suppress the time evolution of a quantum system. This is because each measurement projects the state vector back to the eigenstate.

To put it in other words again, it causes the wavefunction to collapse to the same eigenstate from which the system wanted to evolve away.

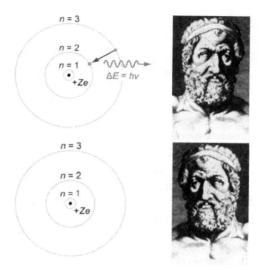

Fig. 147 The Zeno effect:
Observing an atom constantly will avoid its radiating energy.

It can be shown that a quantum system in some specific quantum eigenstate has a probability of 'surviving', called the 'survival probability' P, after some time T and some number n of measurements given as:

$$P = [1 - \Delta H^2 \left(\frac{T}{n}\right)^2],$$

with ΔH a constant that is a function of the Hamiltonian operator and the eigenstate. From this formula, you can see that the probability that the system will be found in the same state depends on how fast one repeats the measurements and that it tends to one, that is, certainty to find it frozen in the same state if T over n goes to zero, meaning a continuous observation.

Of course, the Zeno effect owes its name to one of the famous Zeno paradoxes, the arrow paradox, of the Greek philosopher Zeno of Elea, who tried to argue that motion is an illusion. He imagined time as a line made of timeless segments. Zeno states that for motion to occur (for instance, an arrow in flight), it must change the position it occupies and in any one (duration-less) instant of time, the arrow is neither moving to where it is nor to where it is not. At every instant of time, no motion is occurring. If everything is motionless at every instant, and if time is composed entirely of instants, one reaches the paradoxical conclusion that motion is impossible.

The similarity to this in the quantum Zeno effect is that if the time intervals of our observations become shorter, nearing the 'timeless time unit', the system will no longer be able to move from where it is. The continuous observation will prevent motion.

Notice again how, by 'measurement' or 'observation', one must intend every sort of interaction that causes the projection of the eigenvector onto the eigenstate, not a human observation (even though, for the sake of simplicity, this is what Fig. 147 might suggest with Zeno's severe look, but let us not take things too seriously!). We tried to make it clear, when discussing the Schrödinger's cat paradox, that the interaction of the molecules of the environment with each other due to thermal agitation are also 'measurements'. In fact, in principle, the interaction between molecules in a gas due to thermal agitation could temporally slow down the energy transition of its constituents, or the frequent interaction of a decaying particle with the water droplets in a bubble chamber could influence the probability of the decaying process. However, calculations show that these sorts of interactions are still not long enough to determine appreciable effects.

Yet an experimental verification of this effect was indeed made, by an American group in Boulder, Colorado. [15] The effect was observed in a transition between two states in Beryllium ions (more precisely, between two hyperfine levels, that is, a small splitting of the atomic energy states occurring due to the interaction with a magnetic field). In Fig. 148 left, this is depicted as the transition between states resulting from to the presence of a magnetic field which splits level 1 into 2 and 3: transition 1→2 and back (1 is the ground state, 2 is one of the split hyperfine levels) or 1→3 and back (again, 1 is the ground state, whereas 3 is the other hyperfine energy level).

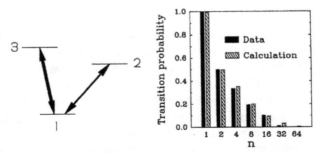

Fig. 148 Experimental proof of the quantum Zeno effect. [15]

The ions were trapped in a device capable of storing charged particles using a homogeneous static magnetic field (called a '*Penning trap*') and were cooled down to near absolute zero temperature (through a laser cooling technique). Short and fast pulses of light, with the frequency with which the transition occurs, conducted the measurement. If the atom was not measured too frequently by the laser beams, it scattered few photons that it absorbed and reemitted. Otherwise, for an increasing frequency, it scattered fewer and fewer photons. Its state can easily be manipulated by radio frequency techniques and, therefore, be prepared in one state or another and then observed in time. The result of the experimental data is shown in the graph of Fig. 148 right. n is the number of measurement pulses of the duration of about 2.4 ms, during a time T of the order 256 ms. The vertical axis is the probability, from zero to one, that the state will undergo a transition. As you can see, there is a dramatic change; the probability that the atoms will undergo a transition becomes much lower whenever we measure its state more frequently. The match between the theoretical data and the experimental data is good.

So, at least in QM, there is much truth in the good old saying that "a watched pot never boils"!

VII. Bell's legacy

1. Is the Moon there when nobody looks?

Here we will go through a short interlude that prepares the philosophical ground for Bell's theorem.

In our analysis of the EPR paradox, we considered the idea that the anti-correlation between the spins that Alice and Bob measure might be predetermined from the outset. That is, we asked whether the situation in which particle A is measured to be in the spin-up state and particle B in the spin-down state (or vice versa) might be a predetermined fact at the time of the production of the entangled particles in the source, not at the instant of measurement. This would, in principle, explain the collapse of the wavefunction as a natural fact. It is simply an abstract mathematical update of the information we have, and no superluminal cause and effect principle would be required.

This idea of thinking in terms of predetermined particle states seems to be a quick and elementary solution to the problem, an assumption which EPR used implicitly. To show that if particles still inherently have a position and momentum (that is, they have simultaneous reality in the sense that it is *"possible to predict it with certainty without disturbing the system")*, they concluded that *"the description of reality given by the wavefunction in quantum mechanics is not complete"*.

After all, it is difficult for every normal human being to think of reality other than objects having definite properties. Whether or not we like the weirdness of the quantum world, we all tend to think of the spin of particles as corresponding to some actual "simultaneous element of reality" that these particles, or whatever kind of objects they are, possess independently of the measurements. This means that objects are supposed to possess definite physical properties with definite values at all times, even if we are not observing it. For technical reasons, we might not be able to measure these properties, but we can't think otherwise than to believe that these exist inherently as part of a reality we call the 'cosmos', the Universe, or existence as a whole—and this is independent of whether or not someone is looking at it. This led Einstein to formulate his famous question to one of his colleagues about the nature of reality: *"Do you really believe that the Moon isn't there when nobody looks?"*

The question of whether things exist independently of one actually seeing them and whether reality is independent of observation is certainly not an entirely new one. Philosophers have long debated such issues. (G. Berkeley's subjective idealism is just one example.) QP, however, gave it new twists and turns, as our assumptions of realism are heavily based on a logic of counterfactual definiteness, which in QP frequently leads to fallacious conclusions.

Moreover, our classical 'Laplacian' notion of reality rests on the assumption that things not only have definite states and properties but that any unpredictable event is random only due to our ignorance of the hidden causes that produce only apparently random phenomena by chance. This is why physicists have long searched for a hidden variable theory that is supposed to explain the randomness of QP such that everything can be predicted, at least in principle, once these variables, their values and the initial conditions of the system are known. In fact, a hidden variable theory presupposes that particles already have definite states, In fact, a hidden variable theory presupposes that particles already have definite states, that is, predetermined properties independent from what we know and before a measurement takes place. From this perspective, a measurement 'reveals' to us only the true state and properties of a particle, which, however, we believe already has existed before and independently from the measurement itself. This is what a hidden variable theory assumes and what our healthy common sense suggests. Some physicists believe that they have found such a theory, which is the De Broglie-Bohm interpretation of QM but so far there is no consensus and no empiric evidence that supports it beyond the status of a mere 'interpretation'.

Experiments like those proposed by EPR were not possible at the time (about 1935). Complex electronic devices and optical instruments like laser technology had yet to arrive on the scene and theoretical research into these philosophical issues had been forgotten for a long time. The foundations of QM, especially those pertaining to philosophical aspects relating to determinism, non-locality, completeness, realism and so forth, are usually not even taught in physics courses in colleges and universities. The standard physicist does not bother much about finding satisfying interpretations of QM. What prevailed, and to a large extent still prevails, is a sort of "shut up and calculate" approach.

Indeed, the abstract theory of QM worked extremely well. With only the mathematical theory, it was possible to make huge advances in particle physics, solid state physics and astrophysics. In addition, practical applications, especially with semiconductors, were so overwhelming—and the industrial and economic growth that resulted from them was so appealing—that very few cared about deep questions on the meaningfulness of the quantum world. After all, Silicon Valley industries or the like look for engineers and physicists who can build new devices that make money. They are not looking for quantum philosophers who are ruminating over 'spooky actions at a distance'. Moreover, between the 1930s-1950s, WWII and the beginning of the cold war inevitably demanded a much more pragmatic mind and approach to science. Plus, as everyone knows, these were the years when physics discovered and realized nuclear fission and fusion, resulting in the first nuclear weapons.

However, even in theoretical physics, the foundations and philosophical implications of QM lost their attraction. This is because, over the years, after Einstein's death, QM as a mathematical theory developed further and indeed could successfully predict a large amount of phenomena that were observed in the laboratory or through astronomical observations, without the need to spend time on interpretational questions. For example, the standard model of particle physics, which is able to describe particle behavior and how they exchange the strong and weak nuclear force, is considered one of the most precisely tested physical theories of all time. This gave most physicists so much self-assuredness in the power of mathematics that those few who were instead more inclined toward understanding the deeper physical meanings of things finally had to get out of the way.

However, the purely pragmatic approach is now showing its limits. After over seven decades of intense research into a theory that unites and generalizes QM with GR, a theory of quantum gravity seems to still be very far from appearing on the horizon. One possible explanation of this failure (though not the only one and not even a point of view with which every physicist would agree) is that this is simply because people were too busy making calculations without understanding what the abstract entities on their papers meant. For example, nowadays you still might find high-ranking physicists who became famous for their great achievements in particle physics but who are still as stuck with respect to the interpretation of the uncertainty principle as Heisenberg was. This is because if one is asked to simply calculate, one does not need any interpretations and just applies Eq. 8 straightforwardly.

Notice that after the 1930s until (roughly) the 1970s, despite many discoveries and extremely successful direct applications of QT, no new principles have been discovered or put forward. Most physicists have developed sometimes very complex theories without advancing new conceptual foundations, much less discovering new fundamental aspects of reality. An army of mathematical physicists calculated and found their abstract symbols to have an incredible power of prediction. They were able to predict and formally express the behavior of Nature, though they always resorted to the fundamental principles established by the previous generations, that is, by Schrödinger, Dirac, Heisenberg, Einstein and others. After the standard model of particle physics, which in its basic building block is a theory of the late 1960s and early 1970s, a plethora of quantum gravity theories followed (like string theory, to mention the most notorious one). However, these could not go beyond a mere theoretical exposition, and so far remain speculations that are still awaiting experimental validation. Even worse, as of the current date (2019), data from particle accelerators seems to indicate that string theory isn't anything in which Nature is interested. Particle physicists are desperately looking for a 'new physics', especially at the Large Hadron Collider (LHC) (which to date is the largest particle collider in the world) in Switzerland, but continue to find nothing. One of the reasons for this is, among other technical ones, is of a sociological and pedagogical nature: Too much emphasis has been placed on the strictly analytical skills of students, researchers and future generations of scientists, and no place was left for those who seek the foundations of the concepts we are using in our theories. (The author devoted an entire book to this specific topic; that book elaborates on this aspect in detail [16].) To get a picture that justifies this state of affairs, you can consult a brief historical timeline of the developments of QP and the SM in Appendix A.II.

At any rate, EPR's paper was almost forgotten and wasn't rediscovered until three decades later. In 1964, when John Stewart Bell wrote a paper entitled *"On the Einstein Podolsky Rosen paradox"* [17], an air of change came. In his free time, Bell was questioning the foundations of QM. (He became much more famous for his unpaid work than for his job at CERN, which may tell us something about how modern science is funded.) Bell investigated how it may be possible to prove or disprove one, or more than one, of the ontological implications that EPR's reasoning implied in QM for locality, completeness and realism.

Fig. 149 John Stewart Bell (1928-1990)

First, he generalized correlations among observables in QM. From that, he obtained an inequality which must be observed by a local, complete and realistic theory. His proof was quite intricate and would need a certain amount of math. We will present it here in a simplified version which nevertheless will convey the essence. To do so, in preparation, we must still develop more technical aspects to fully grasp the implication of Bell's theorem.

2. Photon entanglement

Towards the end of the section on spinors and photon polarization, we saw how the Malus law leads to the respective probability law, which tells us about the probability that a single photon will pass through a polarizer oriented along a direction with an angle θ relative to the polarization axis of linearly polarized light, the squared cosine function law: $p(\theta) = cos^2(\theta)$.

This law was applied to one source and one polarizer. The question, however, is: Does it also hold for entangled particles? After all, these are particles which seem to be in quite a different physical state than non-entangled particles. Moreover, so far we have focused our attention on experimental setups which measure the singlets, that is, fermions, in perfect anti-correlation. It is time to extend this reasoning to photons, that is, to linear or circular polarization states, and examine how things play out if both Alice and Bob are free to orient their polarizers along every direction, as well as to see whether Malus' law remains valid.

First, let us see what characterizes entangled photons and in what way they differ from entangled material particles such as fermions. Entangled photons can be produced using different techniques. One of the most common is to use a quantum optical phenomenon called '*spontaneous parametric down-conversion*' (SPDC). A non-linear optical crystal (for example, BBO) splits photon beams generated by an incident pump laser into pairs of entangled photons with fixed phase, correlated or anti-correlated polarizations, but half of the wavelength of the original wavelength of the incident laser beam. In this down-conversion, of course, the energy and momenta of the photons are conserved. (The single incident photon with some frequency has the same energy as the sum of the energy of the outgoing photons with half the frequency; just think of Planck's equation for the photon energy.) Two polarization types play an important role.

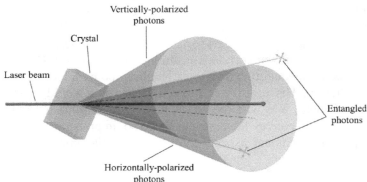

Fig. 150 Creation of type-II entangled photons by SPDC with a BBO crystal.

The type-I SPDC entangled photons is that in which the two photons turn out to have the same polarization — that is, both are vertically-polarized or both horizontally-polarized. These can be described by the state vector:

Type-I SPDC entangled photons: $\quad |\Psi\rangle = \dfrac{|V\rangle_A\,|V\rangle_B + |H\rangle_A\,|H\rangle_B}{\sqrt{2}},$ $\quad$ *Eq. 38*

which implies that, when doing a measurement, one ends up with two vertically-polarized or two horizontally-polarized photons. The analogy with spin-½ must be taken with care: Their intrinsic angular momentum must always be anti-correlated simply because of the law of momentum conservation, which tells us that the sum of the spins must always be zero. Whereas, no such law exists for entangled photons. There is no 'polarization conservation law' and type-I down-conversion is in line with physical laws.

The type-II SPDC entangled photons is that in which the two photons have orthogonal polarization — that is, one is vertically-polarized, while the other is horizontally-polarized. As an entangled electron singlet is anti-correlated in the spins (180° spin direction difference, always), so are the entangled photons anti-correlated in their polarizations (90° polarization direction difference). Instead of spin-up and spin-down, $|+\rangle$ and $|-\rangle$, of entangled fermions A and B, here it is about the vertical and horizontal polarization, $|V\rangle$ and $|H\rangle$, of entangled photons A and B. However, the essence of the physical phenomenon is the same (even though, one should not confuse the photon's polarization with its spin). This can be translated into the entangled photon state vector in direct analogy to *Eq. 35* as:

Type-II SPDC entangled photons: $\quad |\Psi\rangle = \dfrac{|V\rangle_A\,|H\rangle_B + |H\rangle_A\,|V\rangle_B}{\sqrt{2}}.$ *Eq. 39*

Again, this is the state description of a single and unique undifferentiated whole without distinguishable parts, as long as no measurement takes place on one or the other side of this 'quantum wholeness'. The photons are in an undefined and indeterminate state with polarization present on both particles. Only once a measurement takes place will state vector reduction occur and we will find, with a 50% chance, photon A having vertical polarization and photon B horizontal polarization (the system collapses to the product state $|V\rangle_A\,|H\rangle_B$) or, vice versa, a 50% chance of photon A having horizontal polarization and photon B vertical polarization (the system collapses to the product state $|H\rangle_A\,|V\rangle_B$).

Notice that for both cases the "+" sign has been chosen, as we are dealing with bosons. As you might guess, the negative sign for fermions has something to do with the "-" sign we encountered in its spinor vector representation. (More on that later.) However, this is true only in the special case in which the two entangled photons emerging from the crystal have the same phase. This is a special case we will maintain for the sake of simplicity.

3. Polarization correlation coefficients

Having an idea at this point of what entangled spin-$\frac{1}{2}$ (fermions) and spin 1 (photons) particles are and how they can be produced and are described in QM, we can proceed further towards what became one of the most important achievements in the foundation of physics: namely, Bell's theorem. However, we still need some technical basics. In particular, we must grasp the concept of correlation applied to the polarization of photons in order to see what it's all about, as the proof of Bell's theorem relies heavily on a photonic experimental setup of the EPR-like measurements.

Let us now become a bit more professional and, instead of talking of human beings who might be affected by subjective preferences such as Alice and Bob, speak of detectors A and B, which 'click' if they detect a photon or don't 'click' if they don't. The detectors are considered perfect—that is, once a photon hits them, they always click and emit an electric signal that can be registered by a storage device (typically a computer or simply an electric discharge counter). 'Perfect' means that they never miss a photon, which is only an idealization, as real electronic devices have limited efficiency, do not always count 100% of incoming photons and are affected by background noise. However, this is a practical issue that should not concern us at this stage.

We will discuss experiments like those depicted in Fig. 151, in which a source emits entangled photons towards their respective polarizers and detectors in opposite directions. Eventually, they are separated by great distances to avoid mutual influence on each other. One polarizer has an orientation along direction 'a', while the other has an orientation along direction 'b'.

First, we orient both the left and right polarizers along the same vertical direction, the y-axis (Fig. 151 a), while later the second polarizer will be tilted along the horizontal direction, the x-axis, (Fig. 151 b). For clarity, in Fig. 152, the two settings are shown from a side-on view in which both directions are superimposed on the polarization plane.

To clear the way for a more sophisticated representation that will follow, let us represent the above two possible measurement outcomes in a more compact manner. Label the clicking of a detector with the symbol "+", and the absence of the detection signal with the symbol "-". (Please don't confuse this with the spin-up and spin-down symbols!)

The photons, if they have passed through the polarizers, are revealed by the detectors which click (+) or do not click (-). Let us label the set of possible signals coming from the two detectors as "+ +", "+ -", "- +" or "- -". Between the two polarizers, say, at half the way from both, there is a coincidence monitor which gives an output of "1" if both detectors click or do not click (that is, if it registers a "+ +" or "- -" outcome), and a "-1" if the measurement outcomes are opposite (that is, if it registers a "+ -" or "- +" outcome).

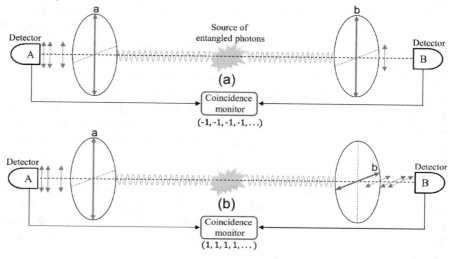

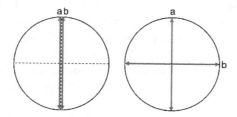

Fig. 151 Experimental setup for the measurement of the polarization states of entangled (type-II) photons with parallel and orthogonal aligned polarizers.

Fig. 152 Superimposed polarization angles for the two orientations of Fig. 151.

Note that we do not diagrammatically show the photons between the source and the polarizers to avoid counterfactual reasoning which would like us to believe that two photons with definite states are flying towards the two polarizers. Instead, we show them only after the measurement, that is, what is observed after the polarizers. Any speculation as to what is there before the measurement is omitted. This will be an important point at the end when we discuss Bell's theorem.

The entangled photons could have correlated type-I polarization, as given by Eq. 38, or a anti-correlated type-II polarization, as given by Eq. 39. Let us first work with type-II polarization photons (which is what Fig. 151

presupposes). At the instant of measurement, at one or the other side A or B, state reduction occurs into product states $|V\rangle_A |H\rangle_B$ or $|H\rangle_A |V\rangle_B$.

This means there are two possible outcomes. The first is that on the left side, a photon with vertical polarization will certainly slip through the vertically oriented polarizer, and detector A will click. Meanwhile, on the opposite side, a horizontally-polarized photon will find the vertically- aligned polarizer and therefore certainly not make it through to detector B, which therefore will not click (a "+ -" pair of signals which will make the coincidence monitor output a "-1"). In the second case, on the left side, a photon with horizontal polarization has no chance of slipping through the vertically oriented polarizer, and detector A will not click. Meanwhile, on the right side, a vertically-polarized photon will find the vertically-aligned polarizer and therefore safely make it through towards detector B, which will click (a "- +" pair of signals which will make the coincidence monitor again output a "-1").

Therefore, when the arrows labelling the photon's polarization on the left side pass through, no photon makes it through the right polarizer, and vice versa. The polarizers' outputs are always opposite to each other's. We say that the detectors' signal is anti-correlated and, hence, the coincidence monitor always gives a -1 answer.

However, keep in mind that as long as the measurement does not take place, they are in an entangled state and nothing can be said about the states of the individual photons. (Again, that's why we did not place a polarization arrow in the pictures between the source and polarizers.)

The two cases can be summarized with the little Table 1 left. For the first case (first row), we put a "+" sign for detector A looking along direction a, and a "-" sign for detector B looking along direction b. Meanwhile, for the second case (in the second row), we put a "-" sign and a "+" sign. (In this case, directions a and b are the same, of course.)

Now, do it all over again with the same polarizer arrangements of Fig. 151a but with the type-I entangled photons given by Eq. 38. In this case, state reduction occurs into product states $|V\rangle_A |V\rangle_B$ or $|H\rangle_A |H\rangle_B$. This means that either on both sides a photon with vertical polarization will make it through and both detectors will click, or on both sides the photons will not make the detectors click because of their horizontal polarization, which will cause them to always be blocked.

This situation is summarized in Table 1 right. Of course, in this latter case, the coincidence monitor will always furnish a "1".

a	b
+	-
-	+

a	b
+	+
-	-

Table 1 Detector responses to type-II (left) and type-I (right) entangled photons for the setup of Fig. 151 a.

We leave it to you as an exercise to verify that, if one repeats the measurements for type-II and type-I entangled photons, but for the polarizer settings of Fig. 151 b, where the polarizers' directions are aligned perpendicularly, one obtains the following outcomes summarized by

Table 2.

a	b
+	+
-	-

a	b
+	-
-	+

Table 2 Detector responses to type-II (left) and type-I (right) entangled photons for the setup of Fig. 151 b.

These were just some examples to make it clear how one can represent the experiments using polarizations.

Now, let us consider a more general situation with the two polarizers pointing towards two generic directions a and b separated by an angle θ, as in Fig. 153. (From now on, for the sake of simplicity, we will omit the photon's polarization arrows after the polarizers and the coincident monitors. However, please always imagine that they are there.)

In this more general case, even if we know that one polarizer lets a photon go through and a detector clicks, we no longer have any certainty about what the other detector will signal. At this point, everything is about probabilities, no longer certainties, because Malus' law gives us a certainty only for the perfectly aligned or misaligned polarizers. For all the other settings in between, we can talk only about the probability that a photon will make it through.

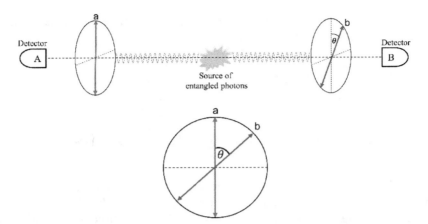

Fig. 153 *Experimental setup for the measurement of entangled photons with polarizers misaligned by an angle θ.*

Intuitively, we can get a feeling as to how things change by varying the angle. Let us proceed by analogy, as we have done in the case of the SG-devices (where we explained how the probability of the spin outcomes varies by smoothly varying the angular tilting of the SG-magnets, see Fig. 80 and Fig. 81). If the misalignment between the two polarizers is small (say, the angle θ is only a few degrees), we have almost the same situation as that of Fig. 151 a. Therefore, in almost all the cases (independent of the case of entangled photons) we will again have the events given by Table 1, apart from only a few exceptions. It is as if polarizer B is 'looking' only a little bit in the horizontal direction and allows, say, for only one over a million photons to make an exception to the fixed outcomes of Table 1. If, however, we tilt the second polarizer (say, by an angle of θ = 45°), much more horizontally-polarized photons are expected to make it through. If Malus' law still holds for entangled photons, the probability of traversing the polarizer is $p(45°) = cos^2(45°) = 0.5$, that is 50%, that both types of polarized photons will traverse the polarizer.

If you repeat the above reasoning compiling the table of events and also allowing for the vertically and horizontally polarized photons to fly through a diagonally oriented polarizer, the list of possible events becomes the following:

a	b
+	+
+	-
-	+
-	-

Table 3 Detector responses entangled photons for the setup of Fig. 153.

That is, all possible combinations of outcomes are allowed. If we maintain one polarizer, say, polarizer A on the left, aligned with the vertical direction and the second pointing at 45° as in Fig. 153, the lines of Table 3 are all equally probable with a 25% chance each.

However, this is a very special situation; generally, the outcomes (the rows of the tables) do not have equal probability. If, for example, for Type-II entangled photons, direction b would shift towards direction a, the first and fourth rows of Table 3 would become less and less probable until they became impossible, once the two directions a and b coincide (that is, the situation returns to that of Fig. 151 a). Table 3 would reduce to Table 1.

Because, in general, each of the possible outcomes has a different 'weight', statisticians prefer to speak of the overall 'correlation' between the outputs of the two polarizers resulting from the measurements instead of the probabilities of each event.

Then, perfect correlation means that both detectors' outputs coincide, that is, either "+ +" or "- -", but never "+ -"or "- +". If they are perfectly anti-correlated, they will always answer in the different, opposite manner, that is, their answer will never coincide; it will be either "+ -" or "- +", but never "+ +" or "- -". If they are not correlated at all, they will answer randomly; all outcomes "+ +", "+ -", "- +" or "- -" are equally probable. To put it in other words, the coincidence monitor is nothing other than a perfect correlation or anti-correlation counter, which, however, gives only a "1" or "-1" correlation, never something in between. But, apart from these special cases, correlation is a real number which varies smoothly from one to minus one.

We could say that, intuitively, the *'polarization correlation coefficient'* C is something which measures the 'sameness' of a collection of quantum binary answers (vertical/horizontal polarization, spin-up/spin-down, photon/no photon, etc.) It could be represented as the probability of registering the same results minus the probability of measuring its opposite, as:

$$C = p(same) - p(different) . \quad Eq. 40$$

If we label as N_{++}, N_{--} the number of coincident clicks of the "++" and "- -" detector signals respectively, and N_{+-}, N_{-+} the number of differing answers for the "+ -" and "- +" signals respectively, the correlation coefficient given by Eq. 40 for a given setting of the polarizers along directions a and b becomes:

$$C(a, b) = \frac{N_{++} + N_{--} - N_{+-} - N_{-+}}{N_{++} + N_{--} + N_{+-} + N_{-+}}.$$ *Eq. 41*

where the 'frequentist' definition of probability has been used, according to which the probability of an event is the number of times it occurred over the total number of events, with the latter as high as possible, ideally infinite.

Simply plug in the former cases and you will see that perfect correlation (the same result, that is, a full match or no mismatch) gives $C(a, b) = 1$, perfect anti-correlation (a different result, that is, a full mismatch or no match) has the value $C(a, b) = -1$, and no correlation with all events is equally probable (50% the same and 50% opposite results, that is, half/half full match or no mismatch) and $C(a, b) = 0$. All the different outcomes in between can give real-number correlations between 1 and -1.

Now that the concept of the polarization correlation coefficient is clear in our minds, we can devote our attention to Bell's achievement.

4. Bell's inequality

In the two preceding sections, we laid down some philosophical and technical foundations for tackling Bell's theorem. If you have not read these sections, please do so, as they are essential for understanding what follows. This will allow us to obtain proof that any classical notion of local realism is incompatible with QM.

There are several ways to illustrate Bell's idea. We won't take his original approach because it would involve more sophisticated mathematical concepts. Bell's original paper is also not necessarily the best approach for intuitively understanding its deep implications. In fact, there are several equivalent proofs of Bell's theorem. We will be inspired by Nick Herbert's version of Bell's proof [18] [19] [20]. Let us consider the following line of reasoning.

First, return briefly to the setting we already saw in Fig. 151 a, in which both polarizers are aligned along the vertical direction. Let us assume that this time the source emits entangled photons of type-I polarization, according to Eq. 38, that is, the photons turn out to be both vertically or both horizontally polarized. Then, as we know, if they have vertical polarization, the two photons always make it through, whereas if the polarization is horizontal, they never emerge from the polarizers. This means that the polarization is perfectly correlated. In this situation, the correlation coefficient is $C(a, b) = 1$ (because $N_{+-} = N_{-+} = 0$ in Eq. 41). The question, then, is: How does this correlation behave if we begin to slowly tilt one of the two polarizers relative to each other?

An instinctive answer might be that even though we are dealing with somewhat special photons (namely, entangled particles), there is no reason to believe that Malus' law no longer holds and one should expect that the law governing the probability that the photons will make it through is still the same squared cosine law. However, a huge problem is attached to that.

Let us first calculate the correlation coefficient which would follow from Malus' law for the photons of type-I polarization. To do that, we must insert Eq. 28 into Eq. 40, which connects the correlation via probabilities as $C = p(same) - p(different)$.

Here, $p(same)$ is the probability of obtaining the same answer from both polarizers ("+ +" or "- -"). In the case of correlated photons, this must be $p(same) = cos^2(\theta)$, with θ the relative angle of misalignment between the two polarizers, as usual. In fact, if both polarizers are aligned, $\theta = 0°$, and the two photons have the same polarizations, there is certainty, $p(same) = 1$, to observe a coincident matching measurement (always "+ +" or "- -"). For the perpendicular polarizers, that is, for $\theta = 90°$, one gets $p(same) = 0$, that is, certainty that both photons will never be detected on both sides (never "+ +" or "- -") because one or the other photon will certainly be blocked, and only one detector will click (always "+ -" or "- +"), that is, 100% mismatch, with no coincident measurements. Meanwhile, for something in between, say, $\theta = 45°$, there is a 50% chance that both detectors will furnish the same answer and a 50% chance that they will differ. All answers, either matching or mismatching, are equally possible.

As to the probability of getting different results, then $p(different) = 1 - cos^2(\theta) = sin^2(\theta)$ simply because it is the remaining probability after having considered p(same).

Therefore, the correlation coefficient for type-I entangled photons, if Malus' law holds, is:

$$C(a, b) = p(same) - p(different) =$$

$$cos^2(\theta) - sin^2(\theta) = cos(2\theta), \qquad Eq.\ 42$$

with the last equality because of an elementary trigonometric relationship. (Look up your good old school tables for trigonometric identities.) We leave it to you to show that for type-II polarized photons, almost the same correlation law holds, with the exception of the positive sign that must be replaced by a negative one. (Recall also that if we took instead of photons singlets of spin half particles, angles are halved; after all, we are discussing an optical analog of an SG-device for spin 1 particles.)

So, the polarizers' response is not given by a linear correlation law but by a cosine function. Let's quickly check it again.

For $\theta = 0°$, $C(a, b) = 1$ → 100% matches (always "+ +" or "- -"). Obvious, because the polarizers are parallel, the two photons' polarization is too. It is a perfect correlation.

For $\theta = 22.5°$, $C(a, b) \approx 0.707$ → 70,7 % matches ("+ +" or "- -"), 29.3% mismatches ("+ -" or "- +"). Most of the detector answers are still the same but a certain number of mismatches begins to appear.

For $\theta = 45°$, $C(a, b) = 0$ → 50% matches, 50% mismatches. On average, the detectors react with an equal number of same results as opposite ones. The data stream is completely random, with no correlation at all.

For $\theta = 67.5°$, $C(a, b) \approx -0.707$ → 29.3% matches, 70.7% mismatches. The average number of mismatches now prevails; the two detectors begin to display an increasingly dissimilar output.

For $\theta = 90°$, $C(a, b) = -1$ → 0% matches, 100% mismatches. The polarizers are perpendicular to each other and when one detects a photon, the other does not (always "+ -" or "- +"). It is perfect anti-correlation.

Now let us take a look at the same situation from a slightly different perspective, taking a more pragmatic approach. Consider the following three scenarios of Fig. 154.

In Fig. 154 a, the right polarizer is tilted clockwise relative to the left one by a small angle δ in direction b.

Still, let us assume that the entangled photons have type-I polarizations, that is, they always have either both vertical or both horizontal polarizations. Due to the tilt in the clockwise direction by δ degrees, the correlation results will be slightly changed from perfect correlation (unity) to, say, for example, $C(a, b) = 0.9$, which means by a net amount of correlation change of 0.1. This means that most of the time, the detectors still agree, except for a few mismatches.

Let us perform exactly the same measurement in the opposite direction b', which is tilted by the same angle δ, but counter-clockwise, as shown in Fig. 154 b. There is no reason to believe that, physically, anything should change: The correlation should again vary by the same amount as it did previously. If one rotates the polarizer in one or the other direction, the situation is perfectly symmetrical and the number of mismatches should be the same, at least on average. (Remember, everything is probabilistic!) Therefore, $C(a, b') = 0.9$ also.

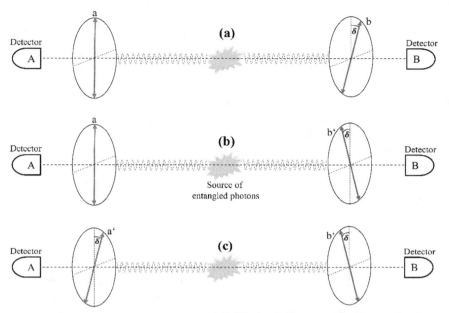

Fig. 154 Polarizers' settings explaining N. Herbert's version of Bell's proof.

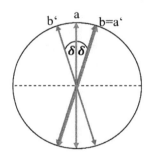

Fig. 155 Polarization diagram for Fig. 154.

Now let us examine the third possibility illustrated in Fig. 154 c. Here we rotate both polarizers by a small angle δ (relative to the upwards direction), one clockwise pointing towards direction a' and the other counter-clockwise pointing towards b'. The relative angular separation between the two polarizers is now two times δ.

The question at this point is: How will the correlation between the two polarizers, which are mutually tilted at a 2δ angle, change compared to the two previous cases with half as much tilt on one polarizer? Or, formally, the question is: How much will $C(a, b) + C(a, b')$ amount to?

One possible answer is that because there can be no difference between tilting the polarizer clockwise or counter-clockwise, the increase in the average number of mismatches induced by doubling the angle must increase in the same manner, that is, at most twice as much. An intuitive perspective by which to see this is to imagine highly correlated (that is, biased) coin-flips. Let us assume a coin-flip almost always produces the same result, say, 99% of the time both heads or both tails. Toss the coin three times. One has a 1% probability of getting a mismatch between the first and second tosses and also a 1% probability of getting the same mismatch between the second and third tosses (already an extreme eventuality). If we repeat this experiment many times, the overall number of mismatches between the first and second tosses plus the number of mismatches between the second and third tosses are together the maximum possible number of mismatches between the first and third tosses, that is 2%. After all, 1+1=2, not 3!

For our purposes, this entails that, given an angle θ that can be subdivided into many small angles δ, say, $\theta = N\delta$ with N being the number of subdivisions of θ (ideally, N tends to infinity), the mismatches observed between the two polarizers could be, at most, N times that for every small angle δ. This means the number of mismatches increases linearly with the increase of the relative angle between the polarizers. This increase in linear mismatch as angle θ grows must then lead to a linear decrease in the correlation coefficient $C(a', b')$ of Eq. 41 from 1 to -1.

But wait a moment! Didn't we say previously that the correlation is given by the cosine function of Eq. 42, which is clearly not linear? To see how the two functions do not overlap even not for small angles, consider the small angle approximation of Eq. 42, which is $C(a', b') = Cos(2\delta) \approx 1 - 2\delta^2$ (with δ in rad). Then, to double the angle δ, the correlation decreases by 4 times, not twice. This contradicts the reasoning above. Therefore, something must have gone wrong here.

One might argue that the idea of the linear growth of the mismatches is not necessarily mandatory. Maybe some hidden physical phenomenon provides for non-linear behavior and would eventually lead to that represented by the cosine function. This would be the classical argument of the 19th-century physicists who would be tempted to reframe and explain the phenomena with a more or less complicated hidden variable theory obeying local classical realism which furnishes the desired result.

The basic assumption that stands behind this is the idea that all the properties of physical objects are well-defined, independently from our knowledge and its measurement. Here, this amounts to saying that the photons' polarization is predetermined and already fixed when they are produced in the light source, though we don't know what it is. According to this local definiteness, the two photons already have well-defined and distinct states during their flight to the polarizers and it is somehow their individual local interaction with the polarizer that determines whether or not the respective detector clicks—and this without being affected by what their 'twin particle' is doing very far away. Therefore, according to this reasoning, perhaps Nature has some mechanism and hidden scheme that allows the two photons to slip or not slip through the two polarizers by producing just the matches or mismatches via the coincident monitor, in accordance with a non-linear correlation law.

This is, however, an argument based on concepts of local realism and that conjectures that there must still be an unknown local theory (eventually governed by hidden variables), which will explain the apparent discrepancy in the perspectives described above. Quantum entanglement does not exist, not to mention 'spooky actions at a distance'; it is just an illusion that will be explained away once we have found a more complete QT. In other words, this local hidden variable theory argument applied to photon polarization is just the EPR argument. They applied it to particles resorting to Heisenberg's uncertainty principle, but it is essentially the same reasoning: QM is not complete and we will sooner or later find a theory that will reconcile it with classical concepts—that is, with a local realistic theory that will also explain why the photon correlation of entangled particles must follow Malus' law.

Is this possible in this context? Could be there, at least in principle, a local theory that explains the non-linearity of the correlation? Bell's clear answer was: No, even not in principle!

Here is the rub. If we seriously stick to a local theory, it must be able to take into account the angular separation between both polarizers which at some point might be separated from each other by light years. The point is that in Fig. 154 c, the photon passing the left polarizer must know instantly the orientation of the right polarizer to display a $\cos(2\delta)$ correlation, while the photon passing the right polarizer must also know instantly the orientation of the left polarizer to behave with the same correlation law. However, both photons were never in contact with the other polarizer; they are produced in the light source in between, and the distance between the polarizers, as well as the distance between the source and the two polarizers, is allowed to have any separation, which means there could be no physical contact among them that allows for any exchange of information about the state and orientation of the relative tilting at the time the photons traverse the polarizers. This, at least, is the case if we want to consider relativity in its present form valid, with no superluminal interactions possible.

So, there is no way around that. If we insist on reality being governed by local phenomena without FTL interactions, we are forced to accept a linear correlation change for the entangled photons. That is, a linear correlation function which is graphically just a straight line connecting $C(a', b') = 1$ for $\theta = 0°$ and $C(a', b') = -1$ for $\theta = 90°$. This is in complete contradiction to Eq. 42, which tells us something a bit different. Fig. 156 draws the two correlation functions. The straight line is that which we obtained by reasoning with a local realistic theory, while the curved line is that of Malus' law which is expected by QM.

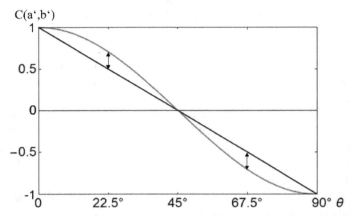

Fig. 156 Correlation functions according to a local realistic theory (straight line) and according to QM (curved line).

It becomes clear that the two lines, which represent two worldviews, those of CP and QM, do not match except for special angles, that is for $0°$, $45°$, and $90°$ of the relative angular separation between the two polarizers. (For larger angles, the same line and curve repeat periodically.) One can also predict for which angles the discrepancy becomes largest: for $\theta = 22.5°$ and $\theta = 67.5°$ (indicated by the two double arrows in the figure).

This is what Bell essentially recognized (even though his original paper adopted a more mathematical approach). It was not just another theoretical speculation or abstract mathematical calculation. It was one of the most profound and decisive insights into the foundations of QP in the last century. Bell not only developed a theoretical approach that highlights a profound divergence between CP and QP but also furnished a very practical method that could be confirmed or disconfirmed by down-to-Earth experiments. It is possible to distinguish empirically with experimentation if we live in a locally realistic Universe, in the sense EPR thought of, or if QM is right and local realism must be ruled out. What had to be done was to produce entangled photons and analyze their behavior by varying the polarizers' orientations, keeping it far enough to not be connected in the relativistic sense, and then check if the coincidence monitor follows a correlation function given by the straight or curved line of Fig. 156. In principle, it is a relatively simple experiment.

To see how this was done, let us dig further into some technicalities. Go back to Fig. 154 c (or Fig. 155) and let us see how things must be written in more general terms of the correlation coefficients from the perspective of local realism. We then know that the number of mismatches that the coincidence monitor will display for the polarizers oriented along some a' and b' directions respectively and separated by an angle $\theta = 2\delta$ must always be less than or equal to the sum of mismatches of the two polarizer settings with smaller angles δ between orientations a,b and a,b'. Instead of the number of mismatches, we can recast this statement in terms of correlation functions. The correlation $C(a', b')$ must always be greater than or equal to the sum of the correlation $C(a, b)$ with $C(a, b')$ (more mismatches → smaller correlation coefficient), which gives a version of the famous *Bell inequality*:

$$C(a', b') \geq C(a, b) + C(a, b') - 1, \qquad \textit{Eq. 43}$$

where we have subtracted a unit on the right-hand side because comparing a sum of correlations with an overall correlation always requires numbers between -1 and 1 (for example, the sum of two perfect correlations is 1+1=2, but it is still a perfect correlation, which must be set to 2-1=1). Keep in mind that this inequality is correct only for average correlations, that is, for the values of the correlation function one obtains after many measurements, and it is only for positive correlation values (that is, for angles between 0° and 90°). We have obtained it in a bit of a heuristic manner; other equivalent and more sophisticated versions exist that have been obtained with more rigorous approaches, but this is more than enough for our introductory purposes.

Eq. 43 is an inequality that must hold if local realism in the sense in which EPR thought of it is correct. If, instead, QM is correct, there must exist some angles for the two polarizers which violate Bell's inequality, as already outlined in the comments on Fig. 156. It is a perfect empiric test that tells us in which reality we live. It is one of the best examples ever of science and philosophy having to come together.

However, for practical purposes, to arrange an experiment which is supposed to prove or disprove this inequality, it is more convenient to build an experimental setup with more generic orientations than that of Fig. 155. For completeness, aside from correlations $C(a, b), C(a, b')$ and $C(a', b')$, there is no impediment to also including correlation $C(a', b)$. Fig. 157 shows the head-on view of the two polarizers for two special orientations. (The 3D view is omitted; it just goes as in the previous cases.) Just these special angles, for 22.5° and 67.5°, have been taken because of Fig. 156, which shows these to be the angles where the violation of Bell's inequality is expected to be the largest.

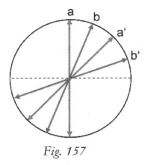

Fig. 157

Then, if you have followed so far, you should also know why, then, Bell's inequality can be extended to the more general one, as:

$$C(a, b') \geq C(a, b) + C(a', b) + C(a', b') - 2.$$

This is a version of the CHSH inequality (CHSH is an acronym for the names of those who introduced it, namely, J. Clauser, M. Horne, A. Shimony, and R. A. Holt) and which was adopted by experimental physicists as the real testbed for Bell's inequality.

The first attempt was made in 1972 by Stuart Freedman and John Clauser, and later with higher statistical accuracy by Alain Aspect, Philippe Grangier, and Gérard Roger in 1982. They measured the linear polarization correlation of the entangled photons emitted by calcium atoms. In Fig. 158 you can see a simple schematic version of the experimental setup.

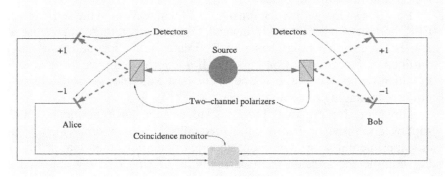

Fig. 158 Experimental setup to test violations of Bell's inequality.

They also used two-channel polarizer splitters, which are able to measure the polarization along two orthogonal directions at once (sort of two polarizers in one) and therefore also allows for the very practical measurement of the two directions a, a' and b, b' at once. This implies the use of four detectors instead of only two which send their signal to the coincidence monitor. The experiment is of a statistical nature. Not just a couple or a few entangled photons have to be measured, but as many as possible (to plug in large Ns in Eq. 41) to obtain probabilities and averaged correlations to test the inequality. By also doing so for several possible polarization angles, so that one can collect a statistically significant amount of data, one can finally test the CHSH inequality, that is, Bell's generalized inequality.

This is easier said than done. Besides the fact that it is not easy to produce entangled photons (and not all the photons created by the BBO crystals we described in Fig. 150 are entangled), one of the main difficulties is maintaining the entangled state during the 'time of flight' to the detectors avoiding decoherence with the environment, the sort we pointed out in the Schrödinger's cat paradox. (This is also one of the main obstacles for building large and efficient quantum computers, as we will see

Fig. 159 Alain Aspect

further on in the upcoming selected advanced topics book edition). In 1964, when Bell's paper was published, the quantum optics technology was still not mature. However, in 1982, the group of Alain Aspect, a French physicist at the University of Paris, carried out an experimental test of Bell's inequalities which was statistically significant and convincingly ruled out any concern for errors induced by the environment or random flukes. [21] Moreover, the photon detectors and all the electronic devices must react extremely rapidly and make the necessary measurements in time lapses short enough to prevent any causal connection between the polarizers and detectors from one side to the other, and which must be as far apart as possible (what in relativity is called 'space-like separation'). This is because one wants to be absolutely sure that one measured photon may not, through some unknown mechanism, send information about its state to the other photon.

However, when technology became mature enough to perform the experiment under these strict conditions, the result was unequivocal: Bell's inequality is, in fact, violated for some angle setting of the polarizers. Experiments show that Nature follows the curved line of Fig. 156, not the straight one. Aspect's experiment was the first in a long series of other experiments that finally confirmed, without any doubt, that Bell's inequalities are violated and that QM is right, whereas a naive conception of local realism, such as EPR desired to preserve, could no longer be considered a viable option.

All this provided strong evidence for QM being a non-local theory. Bell's inequalities and their experimental tests are still debated among physicists. There is still no universal and complete agreement on the subject. Several have tried to revise, elaborate, and reconsider EPR's reasoning, and have tried additional experimental tests of any conceivable sort to find an alternative explanation that saves locality. Several physicists have looked for possible 'loopholes' in Bell's argument and doubted that the experiments possessed 100% certainty being perfectly error-free without possible faults in the experimental setup. That's why these experiments continue to be refined with the advance in technology. The experiments were repeated with photons also several kilometers apart [22], with ultra-fast polarizer settings and events registered completely independently – compared only after the measurements were taken [23] –confirming Bell's inequality violation for electrons and with much fewer statistical uncertainties [24]. The evidence showing how QM violates Bell's inequality has now become overwhelming. Also, other proposed tests which attempted to restore some form of local realism were dismissed by experimental evidence (or, as Karl Popper would have said, they were 'falsified'). Therefore, most physicists regard this issue as settled once and for all. QM is a complete theory that is non-local by its very nature.

Finally, the EPR argument which was looking for a 'bug' in the QT fell in disgrace. This also implies that Nature prohibits any attempt to measure incompatible observables, even in principle. Like it or not, Nature resorts indeed to instant 'spooky actions at a distance' to preserve the principle of complementarity. Einstein would certainly not have appreciated this verdict, but facts have shown that he was wrong.

Despite all these enormous pieces of evidence, since nobody knows what kind of new vision and interpretation of reality should be adopted to replace the old one, the classical reductionists and the Newtonian approach remains the prevailing one. However, we will describe additional examples which finally show how our notions of reality, space and time on which our modern scientific conceptions are based, in addition to being based on a reductionist mechanical and deterministic understanding, turns out to be a very naive world view.

5. Bell's theorem: what is reality?

We can finally draw some conclusions from the experiments that tested Bell's inequality. We have seen that every theory that is local in the sense described by EPR must respect Bell's inequality over all the possible combinations of the polarizers' orientations. The results show otherwise. So, let us summarize the implications of this result.

The violation of Bell's inequality implies that the measurements made at one polarizer cannot depend on the setting of the other one. Photons don't 'know' what to expect or 'plan ahead' because we can conceive of them as far apart such that any physical causal contact is not allowed, at least if relativity holds. Relativity holds because it is a well-tested and experimentally confirmed theory (despite its name of 'theory').

We cannot think of entangled photons, or of any entangled particles, as two objects with their states predetermined in advance at the time of creation and that are already 'pre-programmed' to go through their respective polarizer, or not. This naive idea – according to which there might be hidden variables that encode in the photons (at the time of their creation in the source) the information that tells them what to do later at the instant of measurement – is inconsistent with the predictions of QT. This is because one can also tilt the polarizers along some angles long after the two entangled photons have been created. If the two polarizers are light years apart, Alice and Bob can even wait years after their creation in the source, and eventually opt for a random polarizer orientation shortly before the photons' arrival.

The idea that 'actual elements of simultaneous reality' exist and that they can be ascribed to properties of a physical object, in the sense that EPR wanted to maintain, is doubtful. Our everyday experience, which suggests that all elements of the reality of physical objects, such as particles, must have definite values (a polarization, spin, position, momentum, etc.) independent from observations, even if we don't know their values, is a reasoning that is heavily based on a counterfactual definiteness, which is, however, in flagrant conflict with QT that tells us otherwise. The properties of microphysical objects come into existence at the instant of measurement, not before. Because, again, as long as we have time to orient the polarizers while the photons are in flight, they are not allowed to already be in some definite eigenstate. Only when their polarization is measured with the detectors after the polarizers, and state vector reduction occurs, does their polarization property come into existence and must be (anti-) correlated with that of the other distant photon polarization.

At this point, QT cannot be considered a local theory. Notice, however, that there is still space for a non-local hidden variables deterministic theory, which, in fact, some physicists developed (notably, the de Broglie-Bohm pilot wave theory). These authors believe that, in principle, there is still some hope for building a theory which could explain the quantum randomness in terms of hidden variables reintroducing a classical determinism, provided that non-locality is incorporated.

At any rate, reality must be non-local, and only two possible options remain:

1) If QM has hidden variables, it must be non-local, that is, superluminal signals are possible, at odds with relativity. Non-local FTL interactions are possible.

2) If QM is non-deterministic, that is, does not have hidden variables, it must be a 'non-realist' theory. The wavefunction instantaneous collapse is, indeed, a correct formal description but the wavefunction as such is not a real physical entity. Non-local causality holds and non-local correlations are at work, but no superluminal information can be sent and still nothing physical is FTL, in accordance with relativity.

However, what is no longer acceptable is what EPR had hoped for: to think of a theory which has hidden variables *and* is a local theory.

What all this tells us is that we must give up at least one of the following concepts.

1) Locality: QM is a non-local theory in the sense that a measurement performed by Alice (Bob) of a particle of an entangled system instantly determines the physical state of the other particle and the measurement result Bob (Alice) will obtain, even if they are light years away from each other; FTL signals are possible and relativity must be modified.

2) Incompleteness: QM is a complete theory in the sense that the state vector completely describes the state of a quantum system. A hidden variable theory which explains the correlation between Alice and Bob is not needed.

3) Realism: Ψ is not a physical entity and therefore does not need any form of locality in the relativistic sense. If we abdicate our present conception of realism, we recover a non-local causality which implies FTL correlations between two systems but with no transmission of information, and therefore maintaining compatibility with the relativistic locality and causality.

Notice that the distinction of non-locality between the first and third cases is subtle. In the first case, real 'anti-Einsteinian' and 'anti-relativistic' FTL interactions among particles is intended; one considers that relativity must somehow not be the final story and that superluminal information exchange may be possible – or, to put it in Bell's words, *"there must be a mechanism whereby the setting of one measurement device can influence the reading of another instrument, however remote. Moreover, the signal involved must propagate instantaneously..."* [17] While the anti-realist non-local conception maintains the impossibility of FTL causal interactions, it admits to superluminal quantum correlations provided they can't be used for sending information.

Bell's theorem can summarize this neatly as follows:

"No physical theory of local realism such as a local hidden variables theory can ever reproduce the correlations between entangled particles as predicted by quantum mechanics."

Bell later added: *"If [a hidden variable theory] is local it will not agree with quantum mechanics, and if it agrees with quantum mechanics it will not be local. This is what the theorem says."* [25]

Bell's theorem is a type of *'no-go theorem'*, that is, a theorem stating that if some logical or physical conditions are met, a particular situation is physically impossible.

It is interesting, however, to see how subtle things are and how difficult it is nevertheless to achieve the absolute certitude that this logic is, finally, the ultimately correct way of interpreting the violation of Bell's inequality. Indeed, an even more extreme view of reality exists and is supposed to expose a *'loophole'* in Bell's theorem, namely, the so-called *'superdeterminism'*.

In a BBC interview, Bell explained:

"There is a way to escape the inference of superluminal speeds and spooky action at a distance. But it involves absolute determinism in the Universe, the complete absence of free will. Suppose the world is super-deterministic, with not just inanimate nature running on behind-the-scenes clockwork, but with our behavior, including our belief that we are free to choose to do one experiment rather than another, absolutely predetermined, including the "decision" by the experimenter to carry out one set of measurements rather than another, the difficulty disappears. There is no need for a faster than light signal to tell particle A what measurement has been carried out on particle B, because the Universe, including particle A, already "knows" what that measurement, and its outcome, will be." [26]

Another way out could be, in principle, the *'many world interpretation'* of H. Everett (which we will also address in the second volume). Suffice it to say that few embrace such a view of reality and that these theories are far from being backed by experimental evidence. Yet, at least from the logical point of view, they are still a viable option that can't be excluded.

However, to the author of this book, these attempts that look for any conceivable 'loophole' that hopes to save the rules of common sense remind me of the followers of Ptolemy who refused to accept the heliocentric system and tried by any means to 'save the appearances', fabricating complicated alternative geocentric theories.

In fact, we will pursue a different route. One of the most important takeaway messages of the violation of Bell's inequality is that in QT, at least for entangled particles and other quantum phenomena that entail coherent superposition states, the result of a quantum experiment does not exist until the measurement is made. (This is also the reason why, contrary to other authors, we refrained from depicting two photons 'flying' in opposite directions towards the polarizers; already this is a misleading representation of facts.) A hidden variable theory presupposes definite states, which implicitly means that the result of a measurement already exists before its actualization. If Nature does not allow for even a non-local hidden variable theory, this would mean that the result of an observation occurs at the instant of the measurement which is ultimately the true cause of quantum randomness and unpredictability. According to QT, non-determinism is, therefore, a fundamental property of Nature whereby the outcomes of quantum events have no cause; they are 'cause-less', they simply happen. The author will furnish additional food for thought that will hopefully convince you that any attempt to reintroduce any form of determinism or locality is simply the desperate struggle of an anthropocentric, mechanistic and reductionist worldview destined to die.

VIII. Quantum ontology reborn

In addition to the EPR paradox, other thought experiments were proposed. However, they could not be verified until the mid-1980s because physicists did not have the advanced technological equipment with which to conduct further investigations. It was only about that time when – in particular, due to the availability of new electro-optical devices and more efficient sensors backed by digital technology – some of those experiments became possible. These technological advances opened the door to a whole new set of research lines. The interest in quantum philosophical issues was reborn.

For all those who have philosophically inclined minds and want to focus on this subject in the future, it is really worth going through some of the experimental tests that have been conducted throughout the last few decades, as they reveal the true face of the quantum world – at least that 'face' that gives rise to ontological reflections about the world's structure. Let us examine, then, some experimental realizations which have emerged since then – those which reproduce, with more sophisticated versions, the double slit experiment, through which more insight can be gained into the philosophical aspects of the foundations of QM.

1. The Mach-Zehnder interferometer

The most common optical device used for optical quantum experiments is the Mach-Zehnder interferometer (MZI). Because you will find it cited throughout the literature and employed in its several possible versions, it might be useful to first take a closer look at this fascinating and relatively simple device.

The MZI is a good alternative to the double slit experimental setup. After all, the double slit is already an interferometric device. From this point of view, the MZI and the double slits can be considered optically equivalent. For illustrative purposes only, here we take a sort of 'hybrid representation' between the path and the wave-like perspective to help visualize the interference effects. In Fig. 160 left, you see yet another possible diagram of the double slit experiment depicting two waves travelling along two different paths through the two slits and towards the screen where they interfere. Meanwhile, the bottom diagram shows the MZI, which is built as follows.

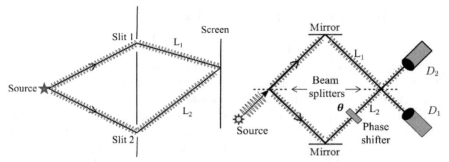

Fig. 160 The double slit (left) compared to the MZI (right).

A source emits EM waves towards a first beam splitter. Nowadays, laser beams are used as light sources because of their coherence – that is, because of their phase stability and also because they are almost perfectly monochromatic due to their narrow wavelength range. Then, as the name suggests, a *'beam splitter'* is an optical device that splits a beam of light into two beams. It is a piece of glass with a special coating (for example, a half-silvered glass plate). The beam splitter sends 50% of the photons in one direction and the other 50% in another perpendicular direction (or any other percentage, depending on the coating). Notice that in doing this, one can obtain two light beams with a perfectly known wavelength and phase and with the same polarization. It would otherwise be almost impossible to create two independent light beams with the same frequency and a stable relative phase and polarization from two independent sources. It is only because we split the very same light beam in two that we do not have to worry about these potential misalignments.

Once the incoming beam has been divided into two beams, these are sent towards two mirrors which reflect them by a 90° angle. The beam travelling along the upper arm of the MZI maintains a fixed optical length L_1. Meanwhile, the lower beam goes through a phase shifter which determines the optical length L_2. A *'phase shifter'* is any device that can delay the phase of a wave by some angle θ. Conventionally, this can be done through the variation of the optical length a beam traverses in a transparent medium, where light travels somewhat slower than in free space according to its refractive index. Therefore, even if the physical length is equal, the variable phase shifter introduces a tunable optical path length difference between the two paths in the interferometer, thereby allowing us to modulate the phase difference between the two beams. This optical path difference, $\delta L = |L_1 - L_2|$, introduces a net phase shift and leads to the interference phenomenon with its black and white fringes (the same δL that we encountered in Young's double slit experiment and that we explained in Fig. 8a). Then, before reaching the detectors, the two beams must go through another beam splitter where they recombine. The waves which went along the upper arm go through the second beam splitter and then towards D_1 or are deflected again towards detector D_2 with a 50% probability for each. The same happens with those photons emerging through the phase shifter. At the second beam splitter, the two beams are superimposed and interfere constructively, destructively or something in between, so that both detectors 'see' the two beams interfering with each other according to some resultant net phase difference.

However, note that while in Young's double slit, the phase between the two beams varies from point to point along the detection screen, in the MZI, the modulation of the phase shifter determines whether the two detectors D_1 and D_2 'see' an incoming wave (the white fringe), no light (the black fringe) or some intensity in between. Moreover, the intensity of the fringes in the double slit has an angular dependence (the fringes become smaller due to the diffraction envelope with growing angles, as in Fig. 8b), whereas in the MZI they repeat themselves with every 180° optical phase shift (see already Fig. 161).

It is, in principle, a very simple interferometric device, and it was already conceived by Ludwig Mach (the son of Ernst Mach, the guy whose name is associated with the speed of sound) and Ludwig Zehnder in 1891. It is somewhat more difficult to build it in practice because one must have perfect and stable control of the optical path lengths on the order of 1/10 of the light's wavelength, which, in visible light, amounts to roughly $50\,nm$ or 1/20.000 mm. (Such small deviations can, for example, already be induced by small temperature changes due to thermal expansion and deformation.)

The interferometric principle that stands behind an MZI is also less trivial than what is usually assumed. This is because what determines the phase difference between the two interfering beams is not just their physical path difference and the optical shift θ of the phase shifter. In practice, one must first consider that the reflection of a light beam by any surface, such as a mirror, also induces another phase shift of 180° each time. Moreover, the beam splitters are also a transparent medium which adds an optical path length depending on its thickness. Complicating things further, it depends on the direction from which a light beam traverses it. The precise description of the MZI should, however, not concern us too much here. (The interested reader can resort to the literature [27].)

Instead, what is essential is that by playing with the phase shifter, constructive or destructive interference between the waves traveling along the two MZI arms takes place and one can switch the beam on and off towards the two detectors as desired: When D_1 'sees' all the light, D_2 does not, or vice-versa. Eventually, the flux can be distributed towards both. The general behaviour of an MZI for any phase shift θ can be obtained by considering the fact that any amplitude of an EM wave can be described by the sine or cosine functions according to the relative phase and recalling that a wave's intensity is given by the modulus square of the amplitude. Then the intensity function in dependence of θ for detector D_1 turns out to be proportional to $sin^2\left(\frac{\theta}{2}\right)$, while that for detector D_2 to $cos^2\left(\frac{\theta}{2}\right)$. The situation is summarized in Fig. 161. The detectors are opposite and switch sides for every integer multiple phase shift of 180° (that is, for $\theta = n\pi$, with n=0,1,2,3…), and share the same intensity for every odd multiple phase shift of 90° (that is, for $\theta = (2n+1)\frac{\pi}{2}$, with n=0,1,2,3…).

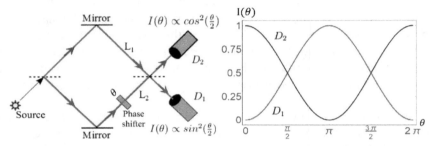

Fig. 161 Response of the two MZI detectors in dependence on the phase shift θ.

The intensity law looks familiar, doesn't it? Yes, it is the good old Malus law we have already encountered with polarizers. This should not come as a surprise: For polarizers, one cuts out a wave by changing the angle of the polarizer, whereas here the same is done using interferometric means.

This is, in principle, how the MZI works. This section was necessary to introduce you to a device that nowadays is ubiquitous among physicists investigating quantum phenomena. However, conceptually, nothing here is new. All this aligns with the wavy nature of light. It is about an EM wave, a light beam that has been split into two waves and that has been recombined somewhere else, such that these two waves can interfere with each other. The question, however, is: What would happen if we sent only one photon at a time through the MZI? And what would happen if we tried to determine through which way it has gone?

2. The 'which-way' experiments

As we have seen in the previous section, the MZI can be used as an alternative testing device to the double slit experiment. As it was in the case of the double slit, here we can perform single photon experiments. This can be done simply by dimming the light source, for instance, using a blackened foil, so that only one photon of energy hν, with ν the frequency of the light, goes through the MZI at a time. As already discussed in the single photon double slit experiment, by 'at a time' we mean that while one photon is 'in flight' through the MZI, no other photon coming from the source is added. If only single photons are detected, detectors no longer provide a continuous output signal but will only 'click' once every time a photon is detected.

So far, we imagined the MZI as being traversed by a continuous flux of EM waves and considered how these waves overlap at the second beam splitter interfering with each other (sort of what we graphically showed in Fig. 160). By dimming the source to such a degree that we can think only of single light particles travelling through the MZI, we can no longer resort to the image of the split wave interfering with itself at the second beam splitter. From the corpuscular point of view, we must think of a single photon taking the upper OR the lower light path, as illustrated in Fig. 162.

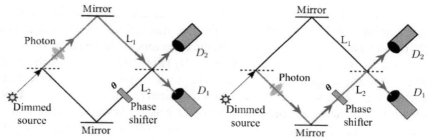

Fig. 162 The 'naive' understanding of the single photon MZI.

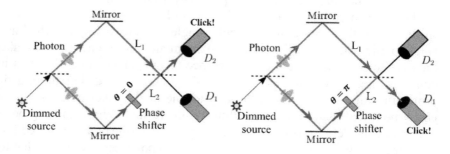

Fig. 163 The less naive understanding of the single photon MZI.

If it must be conceived of as a particle, it can't take both paths, right? The problem with this is that, if so, the photon going either upwards or downwards no longer has anything to interfere with at the second beam splitter. Therefore, taking this naive perspective, we would expect that, below some intensity and photon flux threshold, the interference phenomenon should begin to disappear. All the photons traveling along the upper MZI arm (Fig. 162 left) or through the lower arm (Fig. 162 right) will cause both detectors to click with equal probability. Most importantly, any modulation of the light phase shifter is supposed to have no effect (or, in other words, one might suspect that Malus' law no longer holds for very low-intensity beams).

However, because you are now familiar with Young's single photon experiment and know that an MZI configuration is an equivalent optical realization, you should be able to guess what really happens instead. Even if only a single photon is allowed to pass through the first beam splitter and each time with a 50% chance of traveling along the upper or lower arm, with the phase shifter having zero phase shifting, only detector D_2 will click and nothing will happen at detector D_1 (Fig. 163 left), as we have learned taking the wave-perspective in the previous section.

It looks like the photon always travels along the lower MZI arm nevertheless or that it takes both ways and interferes with itself at the second beam splitter and constructive interference determines its way towards D_2. If you add a $\theta = \pi$ phase shift, again, the detectors exchange their behaviour; detector D_1 will click and never detector D_2 (Fig. 163 right), as if the photon always takes the upper path or it takes both ways and interferes with itself at the second beam splitter and constructive interference determines its way towards D_1. Therefore, the interference effects do not at all stem from the fact that waves or photons travel through the MZI arms. The dimming does not make any difference; only the phase shifter determines where the photons will be detected and which path they seem to take.

The same also fits for all the other possible configurations of the phase shifter. Even though we observe only single photons being absorbed as a single particle at a time, that is, as a quantum amount of energy that makes the detectors 'click' once, the interference phenomena can be observed by counting these clicks for every phase shift angle θ and determining the probability that they will slip through to one or the other detector. It can easily be tested that all the same things with variable phase shifting occur when we considered EM waves, according to the cosine or sine squared law. As in the case with the photon counting through polarizers, we know that for single particles, the Malus intensity law is nothing other than a probability law. If we set the phase shifter at an angle θ the probability that the photon will make detector D_1 click is $\sin^2\left(\frac{\theta}{2}\right)$ and the probability that it will make detector D_2 click is $\cos^2\left(\frac{\theta}{2}\right)$. That is, in Fig. 161 we can replace the intensity function $I(\theta)$ with a probability function $p(\theta)$.

Therefore, as is so typical in QM, technically and mathematically everything is clear but conceptually there remains the usual uneasiness. This is because we are forced to think of point-like particles going along both arms and interfering with themselves at the second beam splitter. In addition, there is another weird aspect: If we want to maintain the picture of only one photon travelling through one or the other MZI arm, we must assume that it knows in advance what the phase shifter state is and 'decide' in advance that it must travel through one or the other path. In fact, if the phase shifter is set to zero, as if there were no phase shifter at all, how does the photon know that it must go through the lower arm to make detector D_2 click? Or, if the phase shifter is set to $\theta = \pi$, how does the photon know it must travel along the upper arm to make detector D_1 click? Before arriving at the phase shifter, how could a photon know its orientation? (How can it even know whether there is a phase shifter at all?) It could equally well have decided to go through the other interferometer path. If we stick with the particle picture, we must conclude that the photon already knows in advance from far away in which state the phase shifter is, and from that it choses how to behave!

The other possible conclusion is the same one that we discussed for the double slits: The particle seems to interfere with itself going through both arms; that is, it behaves like a wave. In the MZI, on one side the photon must go through the phase shifter in the lower arm to 'know' its configuration, while on the other side it must also go through the upper arm to determine the interference at the second splitter, which otherwise, in the absence of interference, would always let through an equal amount of light towards both detectors. This is analogous to the double slit, in which we saw that, also there, if you send single photons one at a time, you will have the single particle spots on the screen but the interference pattern as a whole nevertheless builds up after a sufficient photon count. Being more radical, we should not even think either of particles or waves in the first place, as they are only mental extrapolations of our minds. We already know that, don't we?

So, we cannot claim to know along which path the photon travelled. As long as we observe interference, we lose track of the particle nature and have no means of knowing which path it takes. This MZI experiment simply shows the very same principle of complementarity between wave and particles, only with a different experimental setup than the double slit experiment.

Now, one might nevertheless try to trick Nature by conceiving of an experiment which forces it to reveal us the path along which the photon travelled in the MZI. A good starting point for doing this could be that of considering the so-called 'which-way' or 'which-path' experiments (from the German 'welcher-Weg'). These are all those kinds of experiments which try to determine which way – that is, which path – a photon (or an electron, or whatever particle) followed inside an interference device. Because this superposition state in which particles are supposed to travel along two or more paths at the same time (recall how Feynman path integrals build upon this idea) is an absolutely counterintuitive physical effect that is hardly acceptable for our Aristotelian, at best Newtonian, mindset. Heisenberg argued that this can be explained away with the microscope analogy. He said that it is the perturbation, the interaction of the microscope with the particle, that causes its uncertainty over position and momentum and destroys the interference pattern. Unfortunately, this point of view is also what you will find in much popular science media. It is even taught in several universities. Yet we saw that this interpretation is wrong; the uncertainty would also be present without any interaction at all. We will find further evidence of this.

In fact, Feynman further extended Heisenberg's microscope analogy. In Fig. 164 left we have the usual double slit experiment seen from the perspective of the wave-particle duality.

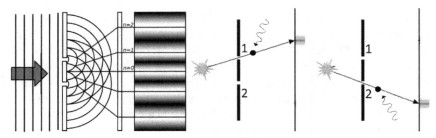

Fig. 164 Left: no information on which-way. Right: the which-way experiment.

Wherever one might find the particle on the screen, one has no means of knowing which way it chose to travel, if it went through the upper or lower slit. To find out, Feynman used Heisenberg's microscope analogy but for a different purpose: not for the determination of the particle's precise position but to establish which way it takes, that is, through which of the two slits it passes, by 'looking', so to speak, very closely at one of two slits using a test photon intended to reveal the particle's 'choice', as shown in Fig. 164 right. (Think of a material particle, like an electron, going through the slits and a photon as the test particle.)

From collecting and analyzing the direction of the scattered test photon, we should be able to establish which slit the particle went through. After all, this is what every radar system does: It emits a stream of radio waves which are reflected by an object (say, an airplane), which are then received by an antenna and through electronic and digital analysis, which measures the time of arrival and the form of the reflected waves, can reconstruct the object's whereabouts. The same principle holds here: By shooting a single photon onto the electron passing through one or the other slit and by measuring the reflected photon direction and energy (recall Compton's scattering process), one should be able, at least in principle, to determine which slit the electron went through. Because Feynman originally proposed this idea (with atoms instead of electrons), it is called '*Feynman's atomic light microscope*'.

What happens then? The fact is, once you are able to establish whether the particle, which appears on the screen later as a spot, has gone through one or the other slit, you will lose the interference pattern! The only attempt to establish the particle's choice of direction will make the interference fringes disappear. The wave-like behavior of the single photon, of this 'something' we call a 'particle', remains as long as we maintain the uncertainty over which way it travelled through.

You might say that this is, of course, quite obvious because, from Heisenberg's point of view, we could easily explain this using the fact that if one detects the particle passing through, say, slit 1, one must inevitably interact with it. And because at the quantum level even the little momentum of a tiny photon is large enough to considerably perturb the system imparting to the particle a kick, a momentum, which is just that uncertainty over the momentum that appears in the uncertainty principle, scattering it away from its path more or less randomly in one or the other direction, it is therefore inevitable that this destroys the interference pattern. However, we said that this interpretation does not stand up to deeper analysis. And, indeed, there is another way to interpret all that. It is not the interaction in and of itself that destroys the interference, the wave-particle duality of photons or matter particles; rather, it is our attempt to force the system into a definite state that selects one of the possible quantum eigenstates and that makes the superposition state disappear.

Speaking more abstractly, the simple attempt to force a system from a superposition to a definite state is usually described as an attempt to gain information about which way – that is, about which slit – the particle went through, and it is precisely this action that destroys the interference pattern. There isn't anything like a particle in the first place; there is a wavefunction $|\Psi\rangle$ evolving in time and space which, after having gone through the two slits but before interacting with the test photon, must be defined by the superposition state:

$$|\Psi\rangle_a = \frac{|e^-\rangle_{S_1} + |e^-\rangle_{S_2}}{\sqrt{2}}$$

with $|e^-\rangle_{S_1}$ and $|e^-\rangle_{S_2}$ representing the electron's eigenstate going through slit S_1 or S_2. When the interaction with the test photon γ occurs it gets entangled with the electron. The overall wavefunction then becomes:

$$|\Psi\rangle_b = \frac{|e^-\rangle_{S_1}|\gamma\rangle + |e^-\rangle_{S_2}|\gamma\rangle}{\sqrt{2}},$$

that is, the test photon interaction still does not collapse the wavefunction but modifies it. This is like the case of polarizers; they 'transform' or in that case 'select' or 'filter' the wavefunction, but do not collapse anything. Only once the electron hits the detector does the wavefunction collapse to one or the other possible event, and then the 'game is over'.

Therefore, even if we could establish the slit through which the particle went by an ideally 'interaction-free measurement', the wave-like behavior of the particles would disappear. For Heisenberg, it was the interaction of the measurement device with the particle that destroys interference; however, according to this alternative interpretation, the mere attempt to force the system to furnish us with information about the position of a particle makes the interference fringes disappear, leaving the normal distribution bell-shaped curve.

These are two conceptually subtly different interpretations of what is going on, but physically they state two very different things. The first interpretation, originally that of Heisenberg, states that in an ideal interaction-free experiment, the interference pattern would be maintained. The second interpretation states that the wave-like behaviour is not connected to any physical interaction between the observed object and the measurement apparatus, but is solely a matter of information retrieval, the 'which-way information'. In fact, notice that no law in QM requires energy or momentum transfer or loss for 'lifting' the superposition in the process $|\Psi\rangle_a \rightarrow |\Psi\rangle_b$.

At first glance, one might be tempted to believe that it is impossible to distinguish between the two cases. This is because an ideal interaction-free measurement of the particle at one of the slits violates the laws of physics, namely Heisenberg's uncertainty principle itself: The minimum amount of energy needed to establish the position of the electron behind a slit is already too large and will scatter the electron because it is imposed by the Planck's constant h, that is universal and immutable. So, the argument becomes circular. Feynman was well aware of this and wrote: *"no one has ever found (or even thought of) a way around the uncertainty principle".* [28]

Later, however, it will be discovered that there is a way out of this. In fact, let us see how one can circumvent this experimental difficulty and nevertheless be able to set up experiments that are equivalent to the double slits and which show that information, in the sense of experimental contextuality, is more fundamental than interaction.

A simple method would be to label the photon when it passes in one or the other arm of the interferometer. For instance, this can be done by polarizing light. As we have seen in the previous chapters, natural light is usually polarized randomly in all directions. However, with a polarizer we can select that part of the light which has a specific orientation of the EM oscillations; we can select a specific polarization eigenstate.

Say we polarize all the incoming photons into the MZI along some direction. Once the light coming from the EM source has been polarized with a polarizer (the rectangle between source and the first beam splitter in Fig. 165), the wave/photons proceed towards the two interferometer arms both having a definite polarization.

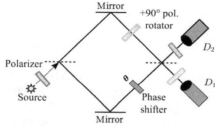

Fig. 165 A MZI which-way experimental setup.

Fig. 166 Working principle of a polarization rotator.

On the lower arm, we have again the phase shifting device (not to be confused with a polarizer; shifting a beam's phase is one thing, while polarizing it is another!). On the upper arm, we introduce a polarizer rotator (the rectangle after the mirror in the upper MZI arm). This is a device which does not select a polarization but rotates the incoming one, say, of 90°; the EM field vectors enter with some angle and are transmitted with a 90° added angle (see Fig. 166). This is the so-called 'Faraday effect', of course, because Michael Faraday, the famous English physicist who made great contributions to electromagnetism, discovered it first. It is an effect that can be obtained by applying a magnetic field to a special transparent medium.

Finally, shortly after the output of the second beam splitter and before the detectors, we insert another polarizer that lets through only that light which has the polarization of the corresponding MZI arm. In front of D_2, the polarizer has the same orientation as that short after the source, whereas in front of D_1 the polarizer is oriented along a 90° tilted direction. By doing so, because we have labeled, with the polarization rotator, the photon in the upper arm to distinguish it from the photon which we imagine went through the lower arm, we have an optical device with which we can control the direction a photon went through the interferometer. In this sense, the test photon of Feynman's thought experiment – which, needless to say, is a measurement that interacts with the particles – has been replaced by the polarization rotator which labels the particle but does so without interacting with it. This is because the polarization rotator, like the polarizers, does not collapse the wavefunction but only modifies or transforms it. (In case you doubt this, read further.)

Consider first the phase shifter set to zero, as if it does not exist. Then we have the following four possible scenarios.

Case I: The photon goes along the upper MZI arm and through the second beam splitter: Detector D_1 clicks because it is a +90° polarized photon and the polarizer in front of D_1 allows it through. D_1 clicks, D_2 does not.

Case II: The photon goes along the upper MZI arm and is reflected by the second beam splitter: The polarizer in front of D_2 blocks it because it is a +90° polarized photon and the polarizer in front of D_2 does not allow it through. Neither D_1 nor D_2 click.

Case III: The photon goes along the lower MZI arm and through the second beam splitter: Detector D_2 clicks because it is a 0° polarized photon and the polarizer in front of D_2 allows it to go through. D_2 clicks, D_1 does not.

Case IV: The photon goes along the lower MZI arm and is reflected by the second beam splitter: The polarizer in front of D_1 blocks it because it is a 0° polarized photon and the polarizer in front of D_1 does not allow it through. Neither D_1 nor D_2 click.

Therefore, half the photons go missing (case II and case IV). Yet this does not concern us because we are not interested in counting photons. Rather, we are interested only in determining which way they travelled through the MZI. The essential point is that we can be sure that when detector D_1 clicks, the photon has taken the upper MZI path. On the other hand, when detector D_2 clicks, the photon has taken the lower MZI path.

So, we have complete control over the 'which-way'.

However, what happens when we modulate the phase shifter? We will then see that nothing happens! And with what probability do the first and second detectors click? We will always measure 50% each, independent of the phase shifter setting. (That is, we recover the naive scenario like that of Fig. 162.) This means that interference is gone and the wave-like property is lost.

This demonstrates that, when we build a measurement device which forces Nature to give us information about which way the photons travel (that would also amount to, say, through which slit in the double slit experiment the photon went through), the interference phenomenon disappears. It is the attempt to retrieve information about the whereabouts of a quantum particle in itself that determines whether or not it will maintain its wave-like properties. As long as we lack information about where the photon is and along which of the two MZI interferometers' arms it has gone through (the MZI without the polarization rotator which 'labels' the photon), interference is preserved and we have the usual interference fringes, as we were accustomed to seeing in the double slit experiment. Otherwise, if we build an experimental setup to force the photon to reveal its whereabouts (the MZI with the polarization rotator which 'labels' the photon), the interference fringes disappear. The wave-particle duality is, therefore, an information-dependent phenomenon. Translating this back to Feynman's microscope, this means that any attempt to detect the photon near one or the other slit to determine which slit the particle has gone through would immediately destroy the interference pattern on the detection screen, independent of how small the perturbation is. In fact, in contrast to Feynman's thought experiment, in the case of the MZI we didn't perturb the photon from its path. We can no longer resort to Heisenberg's explanation who thought that all that happens due to a perturbation of the measurement apparatus with the system to be measured. The wave-like behavior also disappears without any sort of interaction that could disturb the particle's position or momentum. It is only the information over the path, the 'which-way' information as such, that destroys the interference pattern.

That's why many popular science books and videos on the internet depict an 'eye' watching the electron going through one or the other slit, which has led to a very misleading interpretation – namely, that which suggests a supposed role of an observer in QP that perturbs a system and that the outcome of the experiments in QP might have something to do with our consciousness or mind. It cannot be overstated that the experiments in QP have nothing to do with the consciousness or the mind of a human being, as, in a real experiment, the 'eye' is only a detector capable of detecting single photons (such as a sensitive CCD camera or a photomultiplier), which we hardly consider 'conscious observers'. The laws of QP did not come into being with human observers but have existed since the Universe existed; throughout all these billions of years, no human mind or individual conscious observer was needed to make it work. Secondly, it has even less to do with 'perturbations' or 'interactions of an observer with the observed system', as so many like to describe it, because here we have illustrated an experiment that does not presuppose any interaction with the photon at all.

We will see a more striking example of this, discussing another interaction-free measurement. However, we can already perform another simple modification to our which-way experiment to see how the interaction between the measurement apparatus and the system to be measured is not the real issue in QP. In fact, one might state that, after all, there could have been some sort of interaction in the polarization rotator which perturbs the photon. Maybe at the microphysical level, some interaction occurs between the photon and atoms or molecules of the polarization rotator such that it destroys the interference pattern.

There are theoretical considerations as well as practical experimental examples that make it clear that this cannot be the case. For the theoretical arguments, first, one should recall that what the polarization rotator does is only a flipping of the electric and magnetic field of the photon. It does not perturb its path. Moreover, a non-perfectly transparent medium like a polarizer or polarization rotator may occasionally absorb photons but then it destroys them entirely and converts them into heat, excited atoms or mechanical vibrations of atomic lattices (as we have learned with the blackbody cavity). However, this does not change its path or impulse; it simply absorbs these photons, which aren't counted anyway.

Still, to be absolutely sure that the ambivalent wave-particle behavior is not due to interactions or disturbances, let us simply rotate back the polarization of the photon in the upper MZI with a second polarization rotator after the first one, like in Fig. 167.

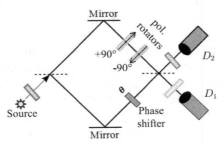

Fig. 167 Erasing the which-way information restores the interference fringes.

The second polarization rotator rotates the EM field back to its original polarization, that is, while the first adds +90°, the second subtracts -90° so that the net polarization remains unaffected relative to the original one that the polarizer set before the first beam splitter. In other words, first we 'mark' the photon but shortly after we strip away its label. This also means that, by 'de-labelling' the photon, we prevent ourselves from again having any chance of knowing which way it went. Photons travelling through the upper or lower arm have, again, the same polarization but no information is left about which path they took.

Therefore, if the disappearance of the interference fringes is due to interaction between the two polarization rotator devices and the photons, no interference phenomenon should be observed because, this time, this supposed 'perturbation' is applied twice. If, however, the disappearance of interference fringes must be ascribed only to the which-way information, we should recover the interference phenomenon.

So, guess what will happen when the experiment is performed? Yes, correct: The interference pattern will nicely turn back. Modulating the phase shifting angle will show how the intensity in detector D_2 goes up and down again. (The light towards D_1 will be dispersed by the polarizer in front of it in the form of heat.) Here, the observational interaction argument à la Heisenberg definitely falls short because even if the two polarization rotators induce any perturbation, the interference pattern – that is, the wave-like nature of light – is preserved.

Note also how, with the second polarization rotator along the upper MZI arm, one 'erases' the information about the polarization status of the photon – that is, one erases the information about the which-way the photon travels. These sorts of experimental set-ups, in which one erases some sort of information over a quantum state, are called *quantum erasers*. The attentive reader will recall how we already saw something of this sort when we used the MSG experimental setup, which erased the information over the spin of the particles. Here, instead of a spin, one erases the polarization label but, in principle, the physical phenomenon is the same. Moreover, notice that we have already used a quantum eraser all the time; it is the second beam splitter which erases the information over the which-way path that the light comes from. Eliminate it and you have the obvious two cases: If detector D_1 clicks, the photon must have gone along the upper arm, while if detector D_2 clicks, it must have gone through the lower one. And, obviously, there could be no interference because there would be no place where the two beams could converge. This is the most trivial example of quantum erasing.

The takeaway message is that we should take superposition seriously by imagining a non-localized 'entity' evolving along both arms and in all possible states. When a photon is sent towards the first beam splitter, it immediately splits in a superposition of states and then is present along both arms of the MZI, as if it is 'gropingly' testing the ground on both paths like a ghostly creature. This 'creature' is, of course, the wavefunction which we must think of not just as a particle or a wave but a state describing an undifferentiated whole. Only when the measurement is done and only when one of the two detectors clicks will the state vector collapse to an eigenstate and the photon seem to have taken one of the two possible paths. Our deterministic world, reducible to particles and parts, becomes intelligible. We may also employ the somewhat imprecise 'which-way' terminology but always keeping in mind that there isn't anything that goes this way or another way; rather, there is only 'something' that goes both ways or, better, something which evolves in a superposition state until it collapses due to a measurement.

Several 'which-way' experiments have been done; this was only one example among many possible. However, all give the same answer. We must conclude that the switch from a wave to particle response depends on the context: If the context is such that it does – or does not – impose a deterministic path along one or the other interferometer arm, we will have a particle or wave-like answer, respectively. Any attempt to imagine a particle as an objective, independent reality in itself, following one or another path independently from our measurement context, fails. It is as if Nature gives us an answer which depends on the context wherein we formulate the question. This is also why QM is said to be 'contextual', in contrast to the 'non-contextual' classical physics. And it is this aspect that is so often misleadingly reported as the supposed 'role of the observer'. There is no such thing.

3. Interaction free experiments: How to detect a bomb without interacting with it

In the previous section, we saw how there is no way to obtain information about the path of a photon that travels inside an MZI without destroying the interference phenomenon. Not only that, we also saw that it seems to 'feel' in advance the presence of the phase shifter, even before reaching it. Of course, this challenges our intuitive naive local realism because this seems to imply that photons along an MZI first 'feel' their way along both arms and 'non-locally see' in advance what they will find on their way but then fall back into one of the arms and yet retain a sort of 'memory' of what they found on the other arm.

A somewhat bizarre application of this was proposed by A. Elitzur and L. Vaidman in 1993 [29], whereby one can build an interaction-free device that tests, with a single photon, the presence of an object on one of the interferometer arms without the photon interacting with

Fig. 168 Avshalom Elitzur and Lev Vaidman

it. One might say that QM allows for 'seeing' an object in complete darkness! The thought experiment is the following.

Suppose you must reveal the presence of a bomb which is extremely sensitive; a single photon hitting it will trigger the explosion. Is it possible to reveal its presence without using any photon at all? Elitzur and Vaidman proposed the use of one of the MZI arms as an interaction-free sensor, as in Fig. 169.

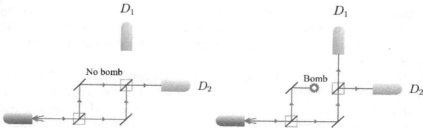

Fig. 169 An interaction free measurement to detect a bomb.

If there is no bomb (Fig. 169 left), we have the simplest version of the MZI (here without phase shifter and polarizers). We have seen that, then, the photon behaves with a wave-like nature, that is, only detector D_2 will always click.

Whereas, if there is a bomb, say, on the upper arm of the MZI, its presence makes the photon behave as a particle. Why? Because by placing a bomb in between, we have constructed a system which allows us to know the which-way of the photon and which destroys the interference; we have a completely different situation than that without the bomb, where interference requires that the photon always be detected at D_2.

In this way (though in a destructive way), we would obtain information about the photon's path and the interference fringes would disappear. In fact, there is a 50% chance that the photon would travel through the arm where the bomb is, make it explode and eventually destroy the bomb tester. There couldn't be a more evident (and destructive) way to signal which way the photon has taken: the upper MZI arm. However, there is also a 50% chance that it would take the other route along the arm with no bomb. Then, when it reaches the second beam splitter, there is again a 50% probability that detector D_2 would click. However, we can't infer, from that, the presence of the bomb because this is also what happens in the case without the bomb. However, there is also another 50% probability that the photon will instead proceed towards detector D_1. In this case, this signals the presence of the bomb, as detector D_1 never clicks without the bomb. So, because in the latter case the photon took the route along the MZI arm, where no bomb is present but which makes the detector click that can't click if the bomb were not there, we have a 25% chance of revealing the presence of the bomb without interacting with it at all!

This is obviously a not-very-recommendable bomb tester because, in the other 50% of cases, the bomb explodes. However, it is nevertheless a nice example of the so-called 'interaction-free measurements' which are possible according to QM. This is because, if we use only one photon and are lucky enough to get that one case out of four in which we reveal a photon making detector D1 click, we can be sure that the bomb is present or any other object on the other arm without having interacted with it. The photon reveals to us the presence of the bomb without interacting with it because it went along another path.

In terms of the wavefunction, the situation can be described as follows. If the bomb is present, that makes the state wavefunction describe the whole system differently than the system without the bomb. If the wavefunction instantly represents the whole system, which also means both of the MZI arms, it must contain the information about the presence of an object. When the single photon goes through the lower MZI arm without the bomb, it 'knows' that it can't interfere with the part of the wavefunction (its 'own itself', so to speak) at the second beam splitter, as this latter has been blocked. And when there is an object on one arm, the photon already instantly knows that an obstruction is on the other MZI arm and it will no longer interfere with itself. Therefore, we must imagine the wavefunction as 'something' that is a transcription of the whole system at once. The wavefunction is, so to speak, an 'information wave about the actual state of being' of the whole apparatus. As long as the measurement is not done, that is, the photon is still on its way towards the detectors, the wavefunction represents the system in a superposition of eigenstates. Only at the time of measurement does it reduce to one or the other eigenstate and make one or the other detector click.

These kinds of thought experiments were, in fact, realized (obviously not with bombs but by using other, less harmful objects like a mirror) and, indeed, confirmed the predictions of QM [30]. The experiments were performed in 1994 by the group of Anton Zeilinger, an Austrian physicist who became particularly known for the experimental verification of the foundations and paradoxes of QM.

Fig. 170 Anton Zeilinger

A couple of years later, a similar experiment was performed by E. H. Marchie van Voorthuysen [31], who presented it live to 500 spectators during a popular science exposition. His account of the reactions speaks volumes. It is worth reading some excerpts from his article: *"A difference between the reactions of professional physicists and others was noticed. Most non physicists were struck by the non-locality of the photon. They saw the problem and wanted the problem to be solved. My statement that this problem probably never will be solved was hardly accepted. Most visitors of 'Majorama 95' had a clear view of science: the remaining problems should be solved in order to make a better world. Now they were confronted with a peculiar problem, but the scientist that presented this problem also declared that the problem is unsolvable! Most physicists were puzzled by the concept of 'interaction-free measurement', apparently because scientists are indoctrinated by the notion that 'a measurement always disturbs the object of the measurement'. This wisdom has become part of our culture. It is known now in many arts and sciences. Physics was taught to be an exception, but QM, where performance of a measurement forces a system to become a member of a set of eigenstates, made clear that the doctrine is valid also in physics. And now it is the same QM that is giving an example to the contrary."*

So, we see that physicists, too, are reluctant to give up the idea that *"a measurement always disturbs the object of the measurement"*. Why are most of them still faithful to the old Heisenberg doctrine, which rests on his microscope experiment but which has been rejected and shown to be false by 80 years of theoretical and empiric evidence? A possible answer might be found in the fact that, after all, these philosophical issues remain a subject of interest only for a minority of scientists (and that can also be inferred in the acknowledgements of the article of van Voorthuysen, in which he had to thank his colleagues and friends who loaned him the equipment to perform the experiment, which would otherwise have been impossible because of the lack of funding for such research). The fact is, even among professional physicists, the Aristotelian way of thinking prevails. Our brains have grown up with a specific set of sensory means and a way of interpreting the world such that we are hardly willing to give them up.

4. The delayed choice experiments

So far, we have seen that Nature apparently responds according to the way we pose the question. Physical reality chooses to give us different answers if we 'ask', so to speak, different 'questions'. If we ask those objects we call 'particles' or 'waves' to have definite properties and follow a deterministic path, we get a corpuscular behavior as our answer, whereas if we renounce to ask about a definite property or path, we get the 'wavy answer' as feedback. We might wonder at this point how exactly does Nature know which question we will ask? When does it decide to answer in one or the other way? Before the particles are going through the double slits? Or before the first beam splitter in the MZI? Or is it when the particles are still traveling towards the screen or detectors? Can we still change our mind, that is, can we change the experimental setup during the flight of the particle?

As we have seen, it looks like photons already know in advance whether they will find one or two slits – or in the case of the MZI, as in the interaction-free measurements in particular, they seem to know in advance whether they will find an active or inactive polarization rotator, a bomb or whatever object. Does the photon, the wave or whatever physical entity that might be traveling in space and time already know in advance what it will find ahead on its path? Or is it able to look into the future and behave accordingly? Is there some sort of retro-causation from the future into the past?

In 1978, these questions prompted the American physicist J. A. Wheeler to design a test that would investigate the true state of affairs. He reasoned mainly with the double slits and proposed the following thought experiment.

Suppose we have the double slit device. Recall that when one shuts one of the two slits, the interference fringes disappear (almost, as far as the single slit's diameter is larger than the wavelength of the photon or the de Broglie wavelength of the material particle; go back to Fig. 51, Fig. 53 and the

Fig. 171 John Archibald Wheeler (1911-2008)

discussion on the pinhole experiment). Wheeler thought the other way around and suggested that, instead of opening and closing one of the two slits, we would use a removable screen behind which two detectors, D_1 and D_2, are placed and that are focused with an appropriate angle towards the corresponding slit 1 and slit 2, and whereby they can determine the slit where the particles appear to go through, as shown in Fig. 172.

This implies that if detector D_1 clicks, this can be due only to a photon that passes slit 1 because it does not 'see' slit 2. Of course, the opposite is true for detector D_2. This arrangement of the detectors allows for the determination of the photon's which-way, and you already know what this implies: No interference fringes are expected. When we use the screen, we already know that we will get the interference pattern (the particles accumulating on the removable screen and which will arrange themselves according to an interference pattern as in Fig. 44 or Fig. 46). However, if we remove the screen and use the detectors instead, we know that we get only particle behavior (Fig. 172, the left and right particles would collect only at the position of the detectors for slit one and two, respectively).

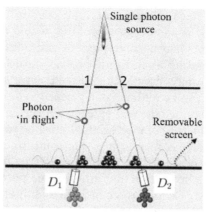

Fig. 172 The delayed choice experiment proposed by J. A. Wheeler.

The point is, according to our human understanding, before the photon passes the slits it must decide whether it must pass through both slits (the case with the screen in place) to interfere with itself and locate itself on an interference fringe, or if it must pass through only one slit (the case with the screen removed and detectors in place) and hit one of the two regions of the lumps of particles. Therefore, it must already know in advance what it will find later – that is, a screen or the two detectors.

Because this apparently 'precognitive' behavior of quantum objects challenges our intuitive understanding of the world, Wheeler wondered what would happen if we chose to send through the slits the photons with the screen in place but, once they passed the two slits and were still in flight towards the screen, one chose to quickly remove it and observe it with the detectors? Would the photons still displace themselves according to an interference pattern? Or, vice versa, what would we observe if we let the single photon go through the slits with the two detectors in place but, while the photon was in flight between the slits and the detector plane, with an extremely fast movement, slide in the screen? Would the photon still hit one of the two spots?

Wheeler's idea was to set up an experiment in which we first let the photon fly through the slits; then, we're still free to change our mind and, by a delayed choice, decide whether to observe it with the screen either still in place or removed. These sorts of quantum experimental tests are called *'delayed choice experiments'* because the method by which we choose to make a measurement is not set in advance but is established while the experiment is going on.

This is a typical example of a thought experiment which is physically possible in principle but unlikely to be realized due to obvious technical difficulties, in this version at least. This is because one must be extremely fast in sliding the detector screen in or out; even if one might be able to do this, the screen itself is a material object that would have to be subjected to extreme accelerations and decelerations in and out of the detectors' plane such that the accelerating force would probably crush it.

However, you may have already recognized how this kind of experimental setup can be realized in a perfectly equivalent manner with an MZI. In fact, we described the equivalence between a double slit experiment and an MZI. Here, instead of a screen that must be removed, it would be the second beam splitter that would play the same role (see Fig. 173). If we remove the second beam splitter, the particle's path becomes known and, obviously, one does not obtain any interference phenomenon. This former configuration, the MZI with the second beam splitter in place, is equivalent to the double slit experiment, while the MZI without the second beam splitter, and behind which the two detectors are placed respectively, is equivalent to the double slit without the detecting screen. In the experimental configuration shown in Fig. 173, the MZI version of Wheeler's delayed choice experiment implies the extremely fast open-closed switching of an ultra-fast second beam splitter which in one state guides light as a beam splitter and in the other state lets light go through without any reflection. The idea is to 'wait' until the photon is 'in flight', say, in the extremely brief time interval during which the photon is traveling along both the MZI arms, and only then decide whether to switch the second beam splitter to open or closed.

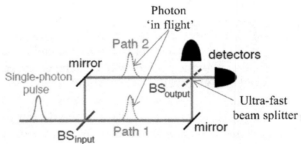

Fig. 173 The MZI version of the delayed choice experiment.

The interesting thing is that modern electro-optical materials exist that are able to switch from being a beam splitter to a normal transparent medium in a few nanoseconds (billionth of a second) without moving anything. Such an ultra-fast beam splitter can be realized with so-called '*Wollastone prisms*'. This allows an experimenter to realize, in practice, Wheeler's delayed choice experiment, which a group of French physicists did in 2007. [32]

Of course, no human being can react so fast and the 'choice' between measuring with either the open or closed configuration was made by fast electronic devices. It is a switch connected to a quantum random number generator which makes the delayed choice randomly. Moreover, placing the second switching beam splitter 48 meters from the first beam splitter allowed the two beam splitters to be sufficiently far away from each other to be space-like separated when the random switching choice was made. Space-like separated means they are causally disconnected in the relativistic sense: The delayed choice is made at the last moment, before the photon enters the second beam splitter, which could have no time to 'inform' the first beam splitter of its state, at least not according to the local realism of relativity, which, as everyone knows, tells us that no information can be delivered faster than light.

So, what was the result? By delaying the choice, with or without the beam splitter (which translates to "with or without the screen" in Wheeler's thought experiment), we will get interference fringes if the second beam splitter is closed during the photon's flight (the beam splitter is optically active, the 'screen' is there) and no interference if it is open (there is no beam splitting, just a transparent medium, no 'screen' present). This is true during any moment until the photon's arrival at the second beam splitter. Therefore, it would appear that the photon indeed already knows in advance, before entering the first beam splitter, what the future state of the second beam splitter will be, and acts accordingly: Going along both MZI arms and interfering or along only one arm and accumulating on one of the two spots! Or, to rephrase this for the double slit, it will behave as if it goes through one or both slits before we decided to detect it with the detectors or the screen respectively. To put it in Einstein's colorful expression, in QM, particles seem not only to interact with 'spooky actions at a distance' but also to possess a sort of 'psychic prescience of the future'. This seems to suggest a sort of quantum retro-causality in which the ordinary causality notion of cause and effect is no longer the same as that in classical physics and that we perceive in our everyday life. Here, the effect seems to precede the cause.

This is one possible interpretation of the facts, held by only few physicists. There is also another way to interpret this without implying exotic 'back-to-the-future' time travel. Simply forget about particles or waves traveling in space; just accept the quantum superposition principle, which does not obey our intuitive understanding of space and time of local realism. Accept the fact that a superposition of states is evolving along both arms and that collapses only at the time of measurement. In fact, in the advanced selected topics book, we will discuss more refined delayed choice and quantum erasure experiments that will furnish several arguments against the idea of quantum retro-causality and which did not find convincing proof remaining, at least so far, making it part of sci-fi, not of QP.

Finally, as pointed out in the which-way experiment of section VIII.2, notice how also in the previous delayed choice experiment, the quick switching of the second ultra-fast beamsplitter from a 'non-beamsplitting' to a 'beamsplitting' state, amounts to an information quantum erasure of the which path information. Inserting the output beamsplitter in Fig. 173 is a which-path quantum erasure procedure, since any photon detected at whatever detector could come from the upper or lower MZI arm, and thereby we loose any knowledge on which path it went through. This is an 'active' quantum erasure type of experiment (the Wollastone prisms is a switching device that can be guided actively by the experimenter or random generator). However, this is a relatively simple quantum eraser, more sophisticated delayed choice quantum erasure versions exist which we will discuss in more details in the second volume.

IX. Concluding remarks and prospects on a second book

I hope that you have been able to follow the intricacies of the quantum world in this first introduction, whose aim was to furnish an extensive overview of the basics of the foundations of QP. This can be taken as an introduction, a sort of self-contained primer.

As you could see, it is certainly not easy to understand the experiments and their results, not to mention the counterintuitive aspects of QM. The real interpretation of the meaning of the quantum phenomena, what they really imply and how we should interpret them in terms of physical models, remains a very controversial matter. There is no general agreement on this issue. These experiments, and especially their interpretations, have baffled and confused the best minds, such as Einstein and Feynman, and they continue to do so today. So, don't worry if you feel as though you didn't entirely grasp the subject.

However, I hope that I was able to convey at least a clear impression of the gigantic conceptual leap that separates classical mechanics, even Einstein's theory of relativity, from the quantum reality. It is my belief that all this mass of empiric evidence should tell us how the nice Laplacian determinism has definitely gone forever, at least in the microscopic world of elementary particles. The deterministic idea in itself of objects having definite properties, such as position, momentum, intrinsic angular momentum, and being a particle or even a wave, is a concept in crisis. There appears only a single undifferentiated whole, a physical 'entity' that, if we might say, is in a 'state of being' and which seems to propagate throughout the measurement set-up in a super position state, yet appears as highly distinct and particularised once observed. The idea of a deterministic Universe reducible to particles or waves, or fields, is only a mental construction of our limited reductionist imagination. These are only emergent properties at the macroscopic scale. They are still Aristotelian, at best Newtonian, notions that emerge in our limited human minds. However, they lose their meaning in the quantum world. Beyond the 'in the mind' reconstructions of the *Homo sapiens*, all these things simply do not exist 'down there' in the sense we like to imagine them.

This is, of course, only a personal assessment of the author, a view that not all physicists share. The worldview and ontology that QP suggests remain a controversial issue. That's why there are so many interpretations of QP. It is also not my intention to furnish a vision and fixed perspective on how the physical world is or must be conceived, as this remains, more than ever, an open question.

With this first introductory course in QP, the intention was not to make you an expert, as that would require a whole set of mathematical and analytic knowledge and skills that is impossible to cover in a few chapters. However, I believe this course has delivered facts that will make you able to understand some of the deep ontological implications of quantum mechanics that, at a university level, are usually burdened with abstractions which frequently tend to obscure the deeper meaning of phenomena or, alternatively, are oversimplified in the popular media with doubtful, if not even pseudo-scientific, arguments. My aim was to furnish an understanding of, and insight into, the subject which is neither too technical, as it is usually taught in universities, nor too sloppy and popular, as it is seen in popular magazines, where misconceptions and misunderstandings frequently creep in. There are also lots of false prophets out there, people who come up with dubious and questionable applications of quantum physics to totally unrelated fields with no scientific grounding and evidence. I hope that you are now in a position, having basic understanding and knowledge, to distinguish and discriminate between pseudo-science, farfetched wild speculations and sound scientific statements on the subject.

This book has covered only the curriculum of the online Udemy course entitled "Quantum Physics: An overview of a weird world". [33] As the introduction already hinted, for readers who are interested in keeping up with the latest developments in the field, it might be useful to remember that a second book is in preparation and will take up where this book has left off. It will cover selected advanced topics such as, to name only a few (in addition to new delayed choice and quantum eraser experiments), the interpretations of QM, the standard model of particle physics and quantum gravity. We will also take a look at quantum teleportation and quantum computing, Bose-Einstein condensates, quantum biology, real or fictitious connections between QP and biology and consciousness, quantum information theory, the black hole information paradox and the holographic principle, quantum cosmology and the Big Bang nucleosynthesis. A couple of concluding chapters at the border between philosophy and metaphysics, such as quantum metaphysics and the quantum woo vs. physicalism, will discuss the tension in which we actually live in our modern cultural age.

Therefore, finally, I hope that I have been able to spark in you more interest and curiosity about the subject so that you feel encouraged to pursue further research on your own and to proceed to the second book. (As already mentioned in the introduction, the interested reader can subscribe to an alert-mail that will inform them about the publication; simply send a note to marco.masi@gmail.com)

And, last but not least, I will be immensely grateful if you post a Reader Review on the book's product page at the online bookstore where you purchased it. These reviews are an essential resource, and it's sad but true that readers seldom bother to post their comments, good or bad! You can also make suggestions in the course of writing a review. Believe me, I read them all. Thank you!

X. Appendix

A I. A little mathematical primer

a. Back to your school algebra!

The very basics you will need are those little algebra skills we were all taught in school, such as being able to deal with fractions, powers and square roots. If you are no longer familiar with them, we need to recapitulate a few things here.

To understand little formulas that involve product terms (like, for instance, that of Heisenberg's uncertainty principle), one should be familiar with the use of algebraic fractions. For instance, given some generic variables x, y, z, from the product $x \cdot y = z$ follows $y = \frac{z}{x}$ or $x = \frac{z}{y}$.

Then, about exponentiation. An exponent is mathematical shorthand that tells us how many times a number is multiplied against itself. In general, given a number b multiplied against itself n times, one writes: $b \cdot b \cdots b = b^n$, with b called the '*base*' and n the '*exponent*'. Note that the base must not necessarily be an integer (such as $2^3 = 8$) but can be any real number. Typical examples are the formulas for the area A of a square and the volume V of a cube, both with side a, given as $A = a^2$ and $V = a^3$, respectively. Don't forget that the square of a negative number always gives a positive one ($(-2)^2 = 4, (-3)^2 = 9, (-4)^2 = 16$, etc.).

This leads us to the square root operation. The square root of a number x is the number y such that $y^2 = x$. One writes: $\sqrt{x} = y$. Examples: $\sqrt{4} = \pm 2, \sqrt{9} = \pm 3, \sqrt{16} = \pm 4, \sqrt{2} = \pm 1.414213562 \dots$, etc.

An important concept is that of the absolute value or modulus |x| of a real number, which is the positive value of a number regardless of its sign. For example, |-3|=3, |-127|=127, |5|=5, etc. The modulus of a real number can be defined as being simply the square root of its square: $|x| = \sqrt{x^2}$.

Also to keep in mind are the elementary laws of algebra. The '*commutative law*' states that the order of the terms in a binary operation does not change the result: a+b=b+a (example: 3+2=2+3=5). This seems obvious but we will see how, in QP, the commutative property is generally no longer true. The '*associative law*' holds when rearranging the parenthesis in expressions containing the same operator does not change the result: $(a+b)+c=a+(b+c)$ or $(a \cdot b) \cdot c = a \cdot (b \cdot c)$. The '*distributive law*' instead holds when two operators are involved: $a \cdot (b + c) = a \cdot b + a \cdot c$. One will frequently find longer variants of this, such as: $a \cdot (b + c + d + e + \cdots.) = a \cdot b + a \cdot c + a \cdot d + a \cdot e \dots$ or, generalizing with indexed terms: $(x_1 + x_2 + \cdots) \cdot (y_1 + y_2 \dots) = x_1 y_1 + x_1 y_2 + \cdots + x_2 y_1 + x_2 y_2 + \cdots$ (the multiplication dot symbol is frequently omitted when it is obvious where it operates). From that follows immediately the basic examples (that lead to the binomial theorem, not listed here): $(x + y)^2 = x^2 + 2xy + y^2$ (if you don't see that, simply take a pencil and a piece of paper and apply the steps given above). Then, one also realizes that: $(x + y)(x - y) = x^2 - y^2$ and should be able to calculate somewhat more complicated expressions like: $(x + 2)^3 = (x + 2)(x + 2)(x + 2) = (x + 2)(x^2 + 4x + 4) = x^3 + 4x^2 + 4x + 2x^2 + 8x + 8 = x^3 + 6x^2 + 12x + 8$.

What about the good old 'Pythagorean theorem'? Everyone in school learns that it is about the relation of the lengths of the sides of a right triangle, namely, the square of the hypothenuse (the side c opposite the right angle) is given by the sum of the square of the other two sides. The Pythagorean equation states: $a^2 + b^2 = c^2$. This is a very important equation that pops up almost everywhere in mathematics, physics, engineering and all their practical applications. For example, it helps establish the distance between two points in space once their coordinates are given, as we will see in the next sections.

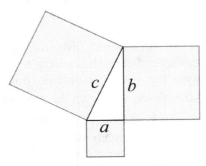

Fig. 174 The Pythagorean theorem

Finally, recall the use of the Greek capital letter Σ ('sigma') for the summation. Some examples: $\sum_{i=1}^{N} i = 1 + 2 + 3 ... + N$; $\sum_{i=1}^{N} i^2 = 1 + 4 + 9 + \cdots + N^2$; $\sum_{i=1}^{N} c_i^2 = c_1^2 + c_2^2 + c_3^2 ... + c_N^2$.

b. Cartesian coordinate systems, scalars and vectors.

Cartesian coordinates are numbers which indicate the location of a point relative to a fixed reference point called the 'origin' of the coordinate system (the intersection between the coordinate axes in Fig. 175). These numbers represent the shortest (perpendicular) distances from two fixed axes (or three planes defined by three fixed axes) which intersect at right angles at the origin. By doing so, one can uniquely identify any point in a two-dimensional (2D) or three-dimensional (3D) space by a set of two or three numerical coordinates, respectively. The 2D Cartesian plane is subdivided into four 'quadrants' and the 3D volume into eight quadrants. Think of it as simply a way to pinpoint objects in space, that is, on a graph or a map. In Fig. 175, for example, the two points P(3,5) and Q(3,0,5) are in the 2D and 3D space, respectively.

In a coordinate system, one can represent more than just points in space, namely, also *'vectors'*. Vectors are quantities given by a collection of numbers and that graphically can be represented by arrows with a length and direction. In physics, lots of physical quantities are vectors (in Fig. 175, for example, the two arrows labeled as $\vec{v}$).

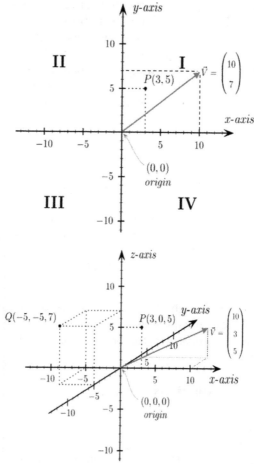

Fig. 175 The 2D and 3D Cartesian coordinate system.

Note how, for vectors, the coordinates are written one over the other, to distinguish them from point coordinates. A practical example is the velocity of an object moving in space and which has a direction and a speed. The speed is given by the length of the vector. The velocity is a vector while the speed is simply a number that tells us how fast the object moves without specifying any direction. This latter kind of quantity is called a '*scalar*'. An acceleration or a force is also a vector, as one needs the direction along which they move or act, too. On the other hand, other scalar quantities could be, for instance, the temperature or the mass of an object. These are just numbers; no directions are needed to specify them.

In general, vectors can be added or subtracted. The resulting vector from the addition of two parallel (or colinear) vectors is simply the sum of its lengths. The resulting vector from the addition of two anti-parallel (opposite) vectors is simply the difference of its lengths. Meanwhile, generally, the resulting vector from the addition of two vectors with generic orientations is done according to the parallelogram law of vector addition. Given two vectors V_1 and V_2, the vector sum $V=V_1+V_2$ is the diagonal of its vector-parallelogram, as shown in Fig. 176.

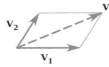

Fig. 176 Parallelogram law of vector addition.

Analytically, this is realized by adding the Cartesian components of each vector to one another. For example, draw it on a piece of grid paper and check by yourself that:

$$\binom{2}{3}+\binom{6}{1}=\binom{8}{4}.$$

This is quite a useful method for summing vectors because all physical quantities, such as velocities, forces, polarizations, EM fields, magnetic momenta and much more, are vectors that can be added according to the parallelogram law.

c. Trigonometric functions and waves

Let us briefly take a look at the formalism that describes waves and that is used extensively in QM. Here, we lay the foundation for the most basic and elementary description of the wave and interference phenomena which are ubiquitous in QP (and in CP as well). This will also open the door to the concept of the wavefunction, a notion of paramount importance in all QP discussions.

You may recall, from your school math classes, some trigonometry and how the cosine and sine functions resemble a periodic oscillation – so to speak, a 'wavy movement.' Be it an EM wave, a sound wave, a water wave or any wavy motions or perturbations, they can easily be described by sine or cosine functions. (Notable exceptions exist, such as shallow water waves, but that will not concern us here.)

The curve going through the origin in Fig. 177 shows the sine function, while the other one is the cosine function.

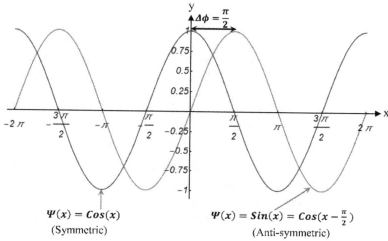

$$\Psi(x) = Cos(x)$$
(Symmetric)

$$\Psi(x) = Sin(x) = Cos(x - \frac{\pi}{2})$$
(Anti-symmetric)

Fig. 177 The sine and cosine functions.

Both have the same domain; that is, they are both traced along the x-axis. This x-axis could represent a spatial, temporal, angular or any other kind of physical magnitude. In this figure, the x-axis represents the angular radians value (radians take values from 0 to 2π instead of the angles in degrees from 0 to 360°) and has been subdivided into steps of $\frac{\pi}{4}$. The magnitude (the 'height') of the wave at a specific value for x, $\psi(x)$, is plotted on the vertical axis. The choice of the Greek letter ψ ('Psi') is not a coincidence; it is the famous and most used letter in all the textbooks on QM. The maximum and minimum magnitudes for both the sine and cosine functions are +1 and -1, respectively.

What distinguishes the two waves is that they are shifted relative to each other along the horizontal x-axis. How much a wave is shifted relative to the other is what determines their *'relative phase'*. The sine function is nothing other than the cosine function shifted by a phase difference of $-\frac{\pi}{2}$ (or the cosine is the sine shifted by $+\frac{\pi}{2}$).

Remember that the concept of phase is always a relative notion, not an absolute one. The phase of a wave must be determined relative to another wave or against a fixed reference. This reference can be another wave or, as it is done usually, relative to an origin of coordinates.

Moreover, an important aspect that differs in the two functions and that will turn out to be useful throughout is that the cosine is a *'symmetric function'*; that is, it is symmetric with respect to reflection along the y-axis (or, its values are equal on the right- and left-hand sides to the origin), whereas the sine function is *'anti-symmetric'*, which means it has a point-symmetry with respect to the origin (or, its values are opposite on the right- and left-hand sides with respect to the origin).

In Fig. 178, we can see a wave described by a cosine function with the x-axis no longer an angle in radians but having the dimension of a spatial quantity (x-axis in meters, cm, etc.) and with a phase shift relative to the spatial origin, that is, with some phase angle ϕ added. This space-dependent function traces a graph of a wave along the space of the x-axis, oscillating a certain number of times per space unit. This latter aspect is accounted for by the so-called *'wave number'*, which is given by $k = \frac{2\pi}{\lambda}$. Therefore, the quantity kx is proportional to a number that tells us how many times a wave of wavelength λ can be arranged into a space interval x, times 2π. (The factor 2π serves the purpose of obtaining an expression given in radians.) Then, the graph of a wave moving along an x-axis in space with initial phase ϕ and wave number k is given by:

$$\psi(x) = \psi_0 \cdot \cos(kx + \phi),$$

where ψ_0 represents the maximum value of the oscillatory movement; it is the *'amplitude'* of the wave (no longer unity, as for the sine and cosine function, but any possible value). Notice that a positive value for the phase ϕ shifts the cosine function backwards, from the right to the left, whereas a negative value results in a shift from the left to the right.

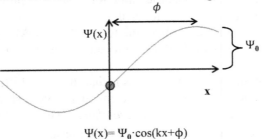

$$\Psi(x) = \Psi_0 \cdot \cos(kx + \phi)$$

Fig. 178 The graph and mathematical expression of a wave in space

However, real waves are not just frozen graphs; they also move with time passing by. To account for the temporal evolution of a traveling wave, another negative phase term which changes with time is added to the description of a wave, making the graph move towards the right direction. (Imagine the dot on the y-axis oscillating up and down and the peaks moving from left to right with time passing by.)

To every periodically oscillating phenomenon, we associate a *'frequency'*, the number of oscillations per time unit. The international SI-base units standard (from the French 'Système International') for the frequency is the *'Hertz'* [Hz], which is the number of oscillations per second, and is labeled with ν (Greek letter 'nu'), for example: $\nu = 50$ Hz means 50 oscillations per second.

One defines $\omega = 2\pi\nu$ ('omega') as the *'angular frequency'*, the frequency (times 2π) with which the wave oscillates. The quantity ωt is proportional to a number that tells us how many times a wave of frequency ν oscillates in the time interval t, times 2π. One might also consider it as a temporal phase shift. As the wavenumber accounts for the spatial coordinates kx, the angular frequency accounts for the temporal displacement ωt. A real wave of amplitude ψ_0, wavenumber k and initial phase ϕ can therefore be represented by a two-valued function:

$$\psi(x,t) = \psi_0 \cdot cos(kx - \omega t + \phi), \qquad Eq.\ 44$$

which depends both on a space coordinate x and a time parameter t. The negative sign in front of the angular frequency assures the wave moves from the left to the right (which is impossible to depict in a still image on paper; just imagine the sin-cos graph moving along the time-axis in a positive direction). With this general expression for a wave, one can describe any physical wavy phenomena in physics. Eq. 44 is the most common mathematical description of oscillatory phenomena.

d. Complex numbers

Let us now say something about complex numbers. Why complex numbers? Well, for the simple reason that they are nowadays used everywhere in science where waves come into play. Moreover, as we shall see, the concept of the wavefunction is a complex valued function. So, let us see first where the idea of complex numbers comes from.

Historically, it was because mathematicians tried to solve equations in which the square root of a negative number could furnish a solution. For example, $\alpha^2 + 1 = 0$. This can be solved only if we admit the existence of a number, called the *'imaginary number'* i, which is the square root of -1, that is, here: $\alpha = i = \sqrt{-1}$. However, there are also other equations, for example, the cubic polynomial equation like this one,

$$a\alpha^3 + b\alpha^2 + c\alpha + d = 0,$$

with a, b, c, d some real coefficients, requesting a solution that is given by a more general complex number, z. In general, any complex number z is the sum of a conventional real part x, Re[z], plus another imaginary part, Im[z], and which is the imaginary number times another real value y:

$$z = x + iy.$$

This means we can represent any complex number graphically on a Cartesian x-y plane, the *'complex plane'*. Here, the real part of any complex number is given on the horizontal x-axis, while the imaginary part is on the y-axis, as shown in Fig. 179.

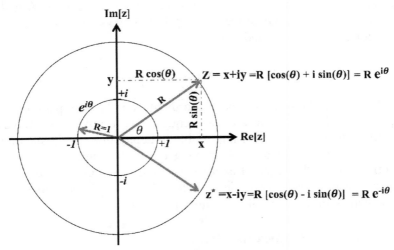

Fig. 179 The complex numbers plane.

Notice that instead of using the Cartesian x-y coordinate axes, it is possible to represent the same complex number with a *'ray-vector'* R and its corresponding angle θ, tracing a circle of radius R, as shown by the (upper) vector **Z** in Fig. 179. Any circle can be represented by the sine and cosine functions, which implies that any complex number can as well. This is because any complex number's real part value along the x-axis can be represented by x= $Re[z]$=R·cos(θ), while its imaginary part value along the y-axis can be represented by y= $Im[z]$=R·sin(θ).

It is customary among physicists and mathematicians to write this in an equivalent but more compact form, using the exponential function, with the

Euler number e=2.718…. The expression $e^{i\theta}$ represents the '*unitary circle*' in the complex plane, the circle of radius R=1, which crosses the x-axis at ± 1 and the y-axis at the imaginary numbers $\pm i$ (that is, when one sets R=1 and $\theta = \pm\frac{\pi}{2}$).

Then one obtains '*Euler's formula*', discovered by the famous Swiss mathematician Leonhard Euler in the 16th century:

$$e^{i\theta} = \cos(\theta) + i\sin(\theta).$$

Putting this together, any complex number can be represented as:

$$z = x + iy = R\cdot[\cos(\theta) + i\sin(\theta)].$$

The important numbers to keep in mind are especially:

$$e^0 = 1$$
$$e^{\pm i\frac{\pi}{2}} = \pm i$$
$$e^{\pm i\pi} = -1.$$

Note that each of these can also represent a $0, \pm\frac{\pi}{2}, \pm i\pi$ ($0°,\pm90°,\pm180°$) phase change. In general, any complex multiplicative factor $e^{i\theta}$ in front of a wavefunction $\Psi(x)$, such as $e^{i\theta}\cdot\Psi(x)$, shifts the wave by an angle θ. For example, $e^{\pm i\pi}\cdot\Psi(x) = -\Psi(x)$ and $e^{\pm i\frac{\pi}{2}}\cdot\Psi(x) = \pm i\cdot\Psi(x)$ are the same wavefunction but phase shifted by $\pm 180°$ and $\pm90°$, respectively.

Therefore, any complex number can also be represented by $z = R\cdot e^{i\theta}$, from which follows:

$$x = Re[R\cdot e^{i\theta}] = R\cdot\cos(\theta), \quad \text{Eq. 45}$$
$$y = Im[R\cdot e^{i\theta}] = R\cdot\sin(\theta).$$

Another concept you will frequently find in the complex number formalism is the '*complex conjugate*'. The complex conjugate is the complex number $z = x + iy$ reflected across the x-axis (the reflection of **Z** into the **Z***ray-vector below the x-axis in Fig. 179). It is obtained only by changing the sign on the imaginary part or on the angle θ as: $z^* = x - iy = Re^{-i\theta}$. The complex conjugate is usually written with a star or bar symbol over z, like z^* or $\bar{z}$. (We will use the former one.)

This is useful to obtain the absolute value of any complex number, the ray vector R. If we define the '*squared modulus*' of a complex number as

$|z|^2 = z \cdot z^*$, and if you do a quick calculation by yourself, you will realize that:

$$|z|^2 = z \cdot z^* = Re^{i\theta} \cdot Re^{-i\theta} = (x + iy) \cdot (x - iy) = x^2 + y^2 =$$
$$= R^2 \cdot [\cos(\theta)^2 + \sin(\theta)^2] = R^2,$$

with the last step because of the term in the square bracket always being unity. This means that the ray-vector R is the modulus of the complex number: R=|z|. You may have noticed how this aligns with the good old Pythagorean theorem (x and y are the legs and R is the hypotenuse of a right triangle).

We shall see why the concept of the squared modulus of a complex number is so important in QM. It will turn out to be the measure for the probability that an event in the quantum world occurs.

Complex numbers are well-suited to represent functions of waves. This is not surprising when we think about how the complex ray-vector of length R, spanning with some angular speed the complex circle, can represent an oscillatory movement. If we project it on the x-axis (just like we did on the y-axis with the dot of Fig. 178 oscillating up and down) from Eq. 45: $x = Re[R \cdot e^{i\theta}] = R \cdot \cos(\theta)$, replace the the ray vector R with the amplitude of a wave ψ_0 (R=ψ_0), while its Euler angle θ is replaced by the spatial and temporal evolution and the initial phase of the wave equation (recall Eq. 44): $\theta = kx - \omega t + \phi$, then the real part of the complex ray-vector is:

$$Re[\psi_0 \cdot e^{i(kx-\omega t+\phi)}] = \psi_0 \cdot \cos(kx - \omega t + \phi),$$

where we have expressed the mathematical formulation of the traveling wave of Eq. 44 with the Euler function.

For compactness, however, physicists and mathematicians describe waves (or any oscillatory and periodic phenomenon that can be described by sine or cosine function) as complex numbers, without taking the real part, simply as:

$$\psi(x, t) = \psi_0 \cdot e^{i(kx-\omega t+\phi)}.$$

This makes calculations easier. Additionally, once a result is obtained, one takes the real part to get an expression in real numbers. The point in using complex numbers to describe waves is that they can retain a phase in space and time with some formal advantages over real trigonometric functions.

So far, we have described the amplitudes and instantaneous field strengths of a wave, which can be a positive or negative number, if its value is above or below the x-axis. This could also be the value of the electric or magnetic field vector of Fig. 4. In this case, the field amplitude can't be taken as a measure of the intensity of a light ray because a negative intensity is not very meaningful. Therefore, what is defined as an intensity of a light beam, that kind of stimulus we can perceive with our eyes, must always be positive and, in physics, corresponds to the squared modulus of the field. For the same reason, a 'squared modulus' of a complex number furnishes the squared ray-vector length. The squared modulus of a complex wavefunction always gives the squared amplitude of the wave,

$$|\psi(x,t)|^2 = \psi(x,t) \cdot \psi(x,t)^* = |\psi_0|^2,$$

and this general quantity is defined as the 'intensity' of the wave independently from any space position, time instant or initial phase.

A final mathematical identity that will turn out to be useful is that of the intensity function resulting from a generic addition of two waves ψ_1 and ψ_2, that is, the square modulus of its sum as:

$$|\psi_1 + \psi_2|^2 = (\psi_1 + \psi_2)(\psi_1^* + \psi_2^*) = \psi_1\psi_1^* + \psi_1\psi_2^* + \psi_2\psi_1^* + \psi_2\psi_2^*$$
$$= |\psi_1|^2 + |\psi_2|^2 + 2\,Re[\psi_1^*\psi_2] \quad \textit{Eq. 46}$$

with the last term because, if we write the complex waves as $\psi_1 = x_1 + iy_1$ and $\psi_2 = x_2 + iy_2$, then:

$$\psi_1\psi_2^* + \psi_1^*\psi_2 = (x_1 + iy_1)(x_2 - iy_2) + (x_1 - iy_1)(x_2 + iy_2) =$$
$$= 2x_1x_2 + 2y_1y_2 = 2\,Re[\psi_1^*\psi_2] = 2\,Re[\psi_1\psi_2^*].$$

This last term will turn out to be the interference term between the two interfering waves or quantum states.

e. Derivative of a function

One of the basic conceptual pillars of calculus is the concept of the derivative of a (real or complex) function. Broadly speaking, it is that quantity which tells us how sensitive to change a function is in its input values. An intuitive way to grasp the essence of the derivative is to think of our everyday notion of speed. We measure the speed of an object as the rate of change of its position in space over time. The function would be, in this example, a position function S in the time parameter t, namely: S(t). The speed v(t) is its derivative because it says something about the rate of change of S in t.

Generally speaking, assuming only real numbers for the sake of simplicity, if we plot the graph of a function f against its variable x in an x-y Cartesian coordinate system such as y=f(x), the derivative of the function measures the rate at which the value y of the function changes with respect to the change of the variable x. It is called the 'derivative' of f with respect to x. Graphically, it represents the slope of the function at each point. Let us take as an example the slope of a linear function. (The graph is just a straight line; see Fig. 180.) In this case, the slope m is defined as:

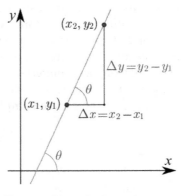

$$m = \frac{change\ in\ y}{change\ in\ x} = \frac{y_2 - y_1}{x_2 - x_1} = \frac{\Delta y}{\Delta x}.$$

Fig. 180 Slope of a linear function.

This can be extended to any function. If we consider the infinitesimal change of a function in y, denoted as dy, for an infinitesimal change in x, denoted as dx, the slope m becomes an infinitesimal quantity as well, and expresses the slope of the function – that is, its derivative – as the gradient (the inclination, the slope of the tangent line in x).

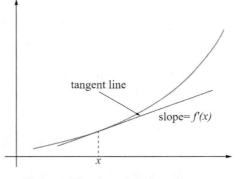

Fig. 181 The slope of a generic function.

In the notation introduced by Leibniz, one denotes the derivative operator (symbol d) as $f'(x) = \frac{df(x)}{dx} = \frac{d}{dx} f(x)$ (read as "f prime of x"). For multiple variables functions, such as f(x,y,z,...), the 'partial derivative' (symbol of the differential operator ∂) of f with respect to x or y or z, etc., is written (chose x) $f'_x(x, y, z, ...) = \frac{\partial f(x)}{\partial x} = \frac{\partial}{\partial x} f(x)$.

Note that one can conceive of the derivative or differential as represented by the action of 'operators' $\frac{d}{dx}$ or $\frac{\partial}{\partial x}$ respectively, that operate on the function f. The operator notion is a central object in QM.

Moreover, of course, it is possible to conceive of multiple derivations (differentiations). The second derivative of a function is the derivative of its derivative, and so on. In general, one writes for the n, the derivative or 'differentiation operator': $\frac{d^n}{dx^n}$ or $\frac{\partial^n}{\partial x^n}$.

So, returning to our previous example, the speed of a body whose movement is described by the function S(t) in space is the derivative $v(t) = \frac{dS(t)}{dt}$. And, by the way, the acceleration a is the second derivative, that is, the rate of change of the speed (think of it): $a(t) = \frac{d^2S(t)}{dt^2}$.

Of course, this gives you only a very superficial and intuitive notion of what it's all about. (College textbooks on calculus are 500-1000 pages thick.) However, for our purposes, this glimpse is sufficient.

f. Integrating functions

A long and articulated introduction to differential and integral calculus would be needed to adequately describe an integral of a given function in mathematics. Here, a not-at-all rigorous and quite sloppy and only intuitive description is given.

Generally speaking, an integral in a 2D space is a measure of the area A that a given function f(x) encloses with the x-axis over an interval [a, b]. Symbolically, it is written as:

$$A = \int_a^b f(x)\ dx$$

Graphically, it represents the shaded area of Fig. 182.

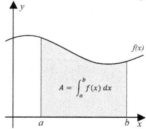

Fig. 182 The concept of the integral: the area under the function and the x-axis.

A similar definition holds for the integration in a 3D space (or, eventually, in an N dimensional space).

This is, however, only a very general understanding of the notion of an integral. The coordinates axes do not necessarily represent space coordinates. For example, the x-axis could represent the flow of time and the y-axis the power consumption of an electric device. The integral over a time interval would then furnish the total energy consumption.

Another practical example which might be useful to introduce us intuitively to the application of integrals in QM could be the calculation of the mass of a body of volume V and density function $\rho(x, y, z)$. The 3D density function tells us how dense the body is at a specific location (x,y,z) in some density unit (say, $\frac{g}{cm^3}$ or $\frac{kg}{m^3}$, etc.). Then, the overall mass of the body is given by the 3D integral over all the volume V of the density function as:

$$M = \iiint_V \rho(x, y, z) \, dx \, dy \, dz$$

What will be of particular interest for our purposes is the integration of the wavefunction; more precisely, its squared-modulus $|\psi(x)|^2$ is a 'probability density' and the 'area' will be, in that case, the probability that a certain quantum phenomenon will occur. More on this in the text.

A II. Physical constant in SI Units

Generally, this book uses the International System of physical Units (SI Units). The SI-base units are the meter (m) for length, the kilogram (kg) for mass, the second (s) for time and the Kelvin (K) for temperature. The SI-derived units for energy are the Joule (J) and the Coulomb (C) for the electric charge. Several other units exist; however, we do not need them in the present treatise. What follows is a list of the physical constants expressed in the SI-Units we used.

Name	Symbol	Value
Speed of light in vacuum	c	$299792458 \, \frac{m}{s}$
Planck's constant	h	$6.626 \times 10^{-34} \text{Js}$
Newton's gravitational constant	G	$6.674 \times 10^{-11} \, \frac{m^3}{kg \, s^2}$
Gravitational acceleration at the Earth's surface	g	$9.81 \, \frac{m}{s^2}$
Boltzmann's constant	k_B	$1.381 \times 10^{-23} \, \frac{J}{K}$
Avogadro's number	N_A	6.022×10^{23}

Electron (and proton) charge	e	$1.602 \times 10^{-19} \mathrm{C}$
Electron's mass	m_e	$9.109 \times 10^{-31} \mathrm{kg}$
Proton's mass	m_p	$1.672 \times 10^{-27} \mathrm{kg}$

A III. Planck's blackbody radiation law

There are several ways to express Planck's radiation law. One can write it in terms of intensity or spectral radiance, or in terms of energy density. Here we chose the latter case:

$$E(v, T) = \frac{8\pi v^2}{c^3} \cdot \frac{hv}{e^{hv/k_B T} - 1},$$

where $E(v, T)$ is the energy density spectrum inside a perfect blackbody cavity at thermal equilibrium at radiation frequency v and at absolute temperature T (h and k_B are the Planck's and Boltzmann's constant, respectively; see A II). Note how, in the denominator of the second term, an exponential function appears which, for increasing frequencies, will increase faster than the square of the frequency itself present in the numerator of the first term. This is precisely what prevents the energy density from diverging into infinity; that is, it prevents the ultraviolet catastrophe.

A IV. Timeline of the developments in QP

This historical timeline is only a subjective selection of events. It is not a rigorous compilation, just the author's personal point of view. However, it should give you an intuitive understanding of how the theory and its practical applications have influenced our understanding of the Universe and world events.

- 1900 Max Planck - quantized energy and blackbody radiation (New principle)
- 1905 Albert Einstein and the photoelectric effect (the photon). (New principle)
- 1909 Hans Geiger, Ernest Marsden, and Ernest Rutherford, scatter alpha particles off a gold foil: atoms have a small, dense, positively charged nucleus.
- 1913 Niels Bohr atomic model based on quantized energy levels.

- 1919 Ernest Rutherford finds the first evidence for a proton.
- 1923 Arthur Compton's photon-particle scattering.
- 1924 Louis de Broglie hypothesis on matter waves. (New principle)
- 1925 Wolfgang Pauli's exclusion principle (New principle)
- 1926 Erwin Schrödinger develops wave mechanics. Max Born's probability interpretation.
- 1927 Werner Heisenberg's uncertainty principle (new principle)
- 1928 Paul Dirac combines QM and special relativity to describe the electron.
- 1930 Wolfgang Pauli suggests the existence of the neutrino.
- 1931 Paul Dirac understands that matter must come in two forms: matter and anti-matter.
- 1933 Enrico Fermi puts forth a theory of beta decay that introduces the weak interaction. Hideki Yukawa describes nuclear interactions.
- 1939 Hans Bethe shows how fusion powers the stars
- 1942 First controlled nuclear fission reaction
- 1945 Hiroshima and Nagasaki
- 1947 Physicists develop quantum electro dynamics (QED) to calculate electromagnetic properties of electrons, positrons, and photons. Introduction of Feynman diagrams.
- 1952 First hydrogen bomb explodes
- 1954 C.N. Yang and Robert Mills develop gauge theories which forms the basis of the Standard Model.
- 1957 Julian Schwinger proposes unification of weak and electromagnetic interactions.
- 1964 Murray Gell-Mann proposes fundamental particles that Gell-Mann names "quarks"
- 1964 John S. Bell proposes an experimental test, "Bell's inequalities," of whether QM provides the most complete possible description of a system.
- 1967 Steven Weinberg puts forth his electroweak model of leptons
- 1974 Stephen Hawking combines quantum field theory with classical general relativity and predicts that black holes radiate through particle emission
- 1980 Alan Guth puts forward the idea of an inflationary phase of the early Universe, before the Big Bang
- 1981 Michael Green and John Schwarz develop superstring theory
- 1982 Alain Aspect carries out an experimental test of Bell's inequalities and confirms the completeness of QM.
- 1983 Carlo Rubbia, Simon van der Meer, find the W^{+-} and Z_0 bosons
- 2012 Discovery of the Higgs boson … and where is supersymmetry?

A V.   Many slits interference and diffraction formula.

Here, the general formula which gives the intensity I of an N-slit reticle as a function of the incidence angle θ and that one can find in every good textbook on optics and electromagnetism. N stands for the number of slits, a stands for the aperture size of the slits, d stands for their common distance from each other, λ is the wavelength, and I_0 stands for the intensity at θ =0. (Eventually, just use normalized units and set I_0=1 to always obtain a maximum at unity value.)

The first square bracket determines the diffraction envelope that 'envelopes' the second, which traces the fringes.

Note how neither the absolute value of the wavelength nor that of the aperture size or the slits' separation determines the interference pattern. Instead, it is their relative size to the wavelength, $\frac{a}{\lambda}$ and $\frac{d}{\lambda}$, that are the determining factors.

$$I\left(\theta\right) = I_0 \left[\frac{\sin\left(\frac{\pi a}{\lambda}\sin\theta\right)}{\left(\frac{\pi a}{\lambda}\sin\theta\right)}\right]^2 \cdot \left[\frac{\sin\left(\frac{N\pi d}{\lambda}\sin\theta\right)}{\sin\left(\frac{\pi d}{\lambda}\sin\theta\right)}\right]^2$$

A VI. Some interesting online resources

Some journals' online portals
PhysicsWolrd: **https://physicsworld.com/**
Nature physics: **https://www.nature.com/nphys/**
Nature latest research: **https://www.nature.com/subjects/physics**

Some interesting blogs
Not even Wrong: **http://www.math.columbia.edu/~woit/wordpress/**
Backreaction: **http://backreaction.blogspot.com/**
Tommaso Dorigo's quantum diaries: *http://blog.darkbuzz.com/*
Sean Carroll's blog: https://www.preposterousuniverse.com/blog/
Shtetl-Optimized: **https://www.scottaaronson.com/blog/**
Dark Buzz: **http://blog.darkbuzz.com/**

Some online forums where to ask questions
(to handle with care!)
PhysicsForum: **https://www.physicsforums.com/**
StackExchange: **https://physics.stackexchange.com/**
Physics Overflow: **https://www.physicsoverflow.org/questions/main**
Quora: **https://www.quora.com/topic/Physics**
Reddit: **https://www.reddit.com/r/Physics**

Other interesting stuff
FQXi Community: **https://fqxi.org/community**
Physics and math applets: **http://www.falstad.com/mathphysics.html**

For the advanced student
Cornell's University physics preprint archives:
https://arxiv.org/list/quant-ph/recent
Philosophy of Physical Science:
https://philpapers.org/browse/philosophy-of-physical-science
Archive for Preprints in Philosophy of Science:
http://philsci-archive.pitt.edu/

XI.   Acronyms

BBO: beta-barium-borate
CM: classical mechanics
CP: classical physics
EM: electromagnetic
EPR: Einstein-Podolsky-Rosen
FTL: faster than light
GR: general relativity
MSG: modified Stern-Gerlach
MZI: Mach-Zehnder interferometer
QFT: quantum field theory
QED: quantum electrodynamics
QM: quantum mechanics
QP : quantum physics
QT : quantum theory
SG : Stern-Gerlach
SM: standard model (of particle physics)
SPDC: spontaneous parametric down-conversion
SR: special relativity
STM: scanning tunneling microscope

XII.   Acknowledgements

I'm particularly grateful to Mark Kelleher, a scientist working in an Antarctic research center, and Tim McGregor, a writer and poet (**http://www.timothyfordmcgregor.com/**) who reviewed this document by making a careful proofread. Without their help this book would have sounded ridiculous!

XIII. Bibliography

[1]   "Physicists can't agree on what the quantum world looks like," *New Scientist,* 4 January 2017.

[2]   A. Comte, The Positive Philosophy, Vols. Book II, Chapter 1, 1842.

[3]   "The official website of the Nobel price," [Online]. Available: https://www.nobelprize.org/nobel_prizes/physics/laureates/1925/.

[4]   in *A Philosophical Essay on Probabilities*, Translated into English from the original French 6th ed. by Truscott, F.W. and Emory, F.L. ed., Dover Publications, 1951.

[5]   J. P. C. e. al., "In search of multipath interference using large molecules," *Science Advances ,* vol. 3, no. 8, 2017.

[6]   D. Bohm, Quantum Theory, vol. 22.11, Prentice-Hall, Englewood Cliffs, 1951.

[7]   H. Rauch and S. A. Werner, Neutron Interferometry: Lessons in Experimental Quantum Mechanics, Wave-Particle Duality, and Entanglement., Oxford University Press, 2015.

[8]   M. B. Albert Einstein, Letter to Max Born; The Born-Einstein Letters, Walker and Company, New York, 1971, 4 December 1926.

[9]   M. B. Albert Einstein, The Born-Einstein Letters, MacMillan, 1971, p. 146.

[10]   A. Pais, Niels Bohrs Times in Physics, Philosophy, and Polity., Oxford: Clarendon, 1991, 1991, pp. 426-427.

[11]   I. Newton, Philosophiae Naturalis Principia Mathematica, General Scholium., 1726.

[12]   T. Maudlin, Philosophy of Physics: Quantum Theory, Princeton University Press, 2019.

[13]   B. P. a. N. R. A. Einstein, "Can Quantum-Mechanical Description of Physical Reality Be Considered Complete?," *Physical Review,* vol. 47, p. 777, 15 May 1935.

[14]   M. Masi, "Video lecture on the Aharonov-Bohm effect," [Online]. Available: https://youtu.be/BoAdLW0DHLw.

[15]   D. J. H. J. J. B. a. D. J. W. Wayne M. Itano, "The quantum Zeno Effect," *Phys. Rev.,* vol. A 41, no. 5, p. 2295, 1990.

[16]   M. Masi, "Free Progress Education", 2017, CreateSpace Independent Publishing Platform - Online: https://www.amazon.com/dp/1539673081.

[17]   J. Bell, "On the Einstein-Podolsky-Rosen paradox," *Physics 1,* pp. 195-200, 1964.

[18]   N. Herbert, "See spot run: a simple proof of Bell's theorem," [Online]. Available: http://quantumtantra.com/bell2.html.

[19]   N. Herbert, *Am. Jour. Phys.,* vol. 43, p. 315, 1975.

[20]   N. Herbert, "How to be in two places at the same time," *New Scientist,* p. 41, 1986, 111.

[21]   P. G. a. G. R. Alain Aspect, "Experimental Realization of Einstein-Podolsky-Rosen-Bohm Gedankenexperiment: A New Violation of Bell's Inequalities," *Phys. Rev. Lett.,* vol. 41, p. 91, 1982.

[22]   J. B. H. Z. N. G. W. Tittel, "Violation of Bell inequalities by photons more than 10 km apart," *Physical Review Letters,* vol. 81, no. 17, pp. 3563-3566 , 1998.

[23]   T. J. C. S. H. W. A. Z. G. Weihs, "Violation of Bell's Inequality under Strict Einstein Locality Conditions," *Physical Review Letters,* vol. 81, no. 23, pp. 5039-5043, 1998.

[24]   B. H. e. al., "Loophole-free Bell inequality violation using electron spins separated by 1.3 kilometres," *Nature Letters,* vol. 526, p. 682–686, 2015.

[25]   J. Bell, Speakable and Unspeakable in Quantum Mechanics, Cambridge University Press, 1987, p. 65.

[26]   J. Bell, Interviewee, *The Ghost in the Atom: A Discussion of the Mysteries of Quantum Physics, by Paul C. W. Davies and Julian R. Brown, pp. 45-46.* [Interview]. 1986/1993.

[27]   S. F. A. R. M. T. K P Zetie, "How does a Mach-Zehnder interferometer work?," *Physics Education,* vol. 35, no. 46, 2000.

[28]   R. Feynman, The Feynman Lectures on Physics. Volume III: Quantum Mechanics, 1956.

[29]   L. V. E. Avshalom C, "Quantum mechanical interaction-free measurements," *Foundations of Physics,* vol. 23, no. 7, p. 987–997, 1993.

[30]   H. W. T. H. A. Z. M. A. K. P. G. Kwiat, "Interaction-free Measurement," *Phys. Rev. Lett.,* vol. 74, no. 24, p. 4763–4766, 1994.

[31]   E. H. d. M. v. Voorthuysen, "Realization of an interaction-free measurement of the presence of an object in a light beam," *American Journal of Physics,* vol. 64, no. 12, pp. 1504-1507., 1996.

[32]   E. W. F. G. F. T. P. A. A. J.-F. R. V.Jacques, "Experimental Realization of Wheeler's Delayed-Choice Gedanken Experiment," *Science,* vol. 315, no. 5814, pp. 996-968, 2007.

[33]   M. Masi, "The Weird World of Quantum Physics," Udemy, [Online]. Available: https://www.udemy.com/quantum-physics/.

[34]   M. Masi, "The Weird World of Quantum Physics," Patreon, [Online]. Available: https://www.patreon.com/quantumphysics/overview.

[35]   B.-G. E. &. H. W. Marian O. Scully, "Quantum optical tests of complementarity," *Nature,* vol. 351, p. 111–116, 1991.

[36]   M. O. T. C. S. P. a. C. H. M. S. P. Walborn, "Double-slit quantum eraser," *Phys. Rev. A,* vol. A, no. 65, p. 033818, 2002.

[37]   Z. K. M. e. al., "To catch and reverse a quantum jump mid-flight," *Nature,* vol. 570, pp. 200-204, 2019.

[38]   X.-S. M. e. al., "Quantum teleportation over 143 kilometres using active feed-forward," *Nature,* vol. 489, p. 269–273, 2012.

[39]   Y.-H. Kim, R. Yu, S. P. Kulik, Y. H. Shih and M. Scully, "A Delayed "Choice" Quantum Eraser," *Physical Review Letters,* vol. 84, no. 1, pp. 1-5, 2000.

XIV.   About the author

Marco Masi was born in 1965 and attended the German School of Milan, Italy. He graduated in physics at the university of Padua, and later obtained a Ph.D. in physics at the university of Trento. He worked as a postdoc researcher in universities in Italy, France, and more recently in Germany, where he worked also as a school teacher for three years.
After he had authored some scientific papers (http://ow.ly/snz6u), his

interests veered towards new forms of individual learning and a new concept of free progress education originated from his activity both as a tutor in several universities and as a high school teacher, but especially from his direct, lived experience of what education should **not** be. From this originated also his desire to write this book on QM which tries to close a gap between the too high-level university textbooks and a too low level popular science approach that is so typical of modern hyped media. He is also interested in metaphysical and philosophical ruminations and loves walking in the woods.

XV. Index

M

N

O

P

CPSIA information can be obtained
at www.ICGtesting.com
Printed in the USA
BVHW041345131019
560778BV00014B/111/P